THE NEW GENESIS

THE GREATEST EXPERIMENT ON EARTH

THE NEW GENESIS

THE GREATEST EXPERIMENT ON EARTH

WOJCIECH K. KULCZYK PhD

The New Genesis Foundation
Frimley, United Kingdom

Publisher:
The New Genesis Foundation
Frimley, United Kingdom
contact@thenewgenesis.foundation

ISBN: 1999906004
ISBN 13: 9781999906009

Cover design: Christopher Harley

To my children,
Katherine and Christopher

Acknowledgements

I wish to thank Dr. Wojciech Brański for reading the manuscript and for many comments which improved the presentation of several difficult issues introduced in this book.

I would like thank Prof Michael Behe and Michael Cremo for permissions to quote material from their books. I acknowledge permissions from Dr Bart Mesuere, Gent University and Eric Clark, National High Magnetic Field Laboratory, Florida State University for reproducing their figures in my book.

I would like to express my gratitude to Witold Koczyński for providing extensive information about the lights in Warsaw in 1983.

Special thanks are due to my daughter Katherine and son Christopher for their comments, preparation of figures and extensive help in editing this book. Without their assistance this book would not reach the publishing stage.

Contents

Introduction

One of the greatest mysteries which man has always wanted to uncover is the explanation for his own existence and the origins of life on Earth. Over the last five thousand years, religion provided an answer to the general satisfaction of most people. The answer was unambiguous – God created everything.

Up until the time of Copernicus it was believed that Earth was the centre of the Universe and that man was the most important being on Earth. The Universe was made for his use to make his life more agreeable. Man had the right to use this world as he saw fit. Since Copernicus' discovery, Earth lost its central position not only in terms of its astronomical location but also in the minds of people. However, man held on to his prime position primarily due to religious teachings which continued to emphasize him as the most important of God's creations.

The situation began to change a few hundred years ago when people started asking specific questions on how the world worked. With the development of a scientific approach on one side and religions loss of authority on the other, interest in the question of the genesis of life came back into focus.

In the 19th century the new scientific paradigm could not accept supernatural origins for life and the creation of man. The interference of God was beyond the limits of scientific thought. This time, science provided more tangible evidence in the form of fossils which showed that life on Earth had existed for far longer than was previously interpreted by religions. Darwin noticed the gradual development of life and formulated his theory of evolution. So evolution took over from religion in the task of explaining the genesis of life on Earth. This new theory provided a very simple explanation. Intelligent life developed as a result of random genetic changes caused by mutations and a process called natural selection which weeded out weaker organisms.

The theory of evolution became very plausible because people could perceive the gradual development of organisms on Earth, from very simple bacteria to man. Although this hypothesis has never been proven, it is treated as true 'by default' because the only other alternative was creationism.

The consequence of the evolutionary theory was that it removed man from his pedestal and treated him as an insignificant, inconsequential product of random mutations. Man not only lost his position but also his purpose in the Universe. At present, a large part of the educated population on Earth, including many religious people, subscribed to this philosophy.

The theory of evolution has been less successful with explaining the origins of life and for about half a century it was believed that life originated from a primordial soup which was subjected to electrical discharges. This hypothesis has been dismissed but even with the proposal of the RNA World and underwater volcanic vents chemistry as the source of the first organisms, the new hypotheses lack any plausible elucidations.

When the evolution hypothesis was proposed our understanding of life, especially the cell, was very limited. It was natural to think that the cell was very simple as it was not possible to penetrate its secrets with an optical microscope. However, with the development of new scientific tools in the second half of the 20th century, a new molecular world which had not been visible before opened before our eyes. Genetics and molecular biology have made huge advancements during the last 60 years providing new and startling knowledge. Our understanding of life has been increasing significantly, especially over the last twenty years and it is still accelerating.

These new discoveries not only show that life has been in existence for almost 4 billion years, but also that right from the beginning it was very complex. Even the simplest single cell organisms, such as bacteria, had very sophisticated structures and control systems. This new information should have prompted evolutionists to reconsider the whole evolutionary approach. But to no avail. What actually happened was the cell's critical evolutionary molecular development was placed in the "dark ages" right at the beginning of life on Earth which lies beyond our investigating horizon. This means that there is no evidence, no fossils, no information at all about this most important period in the development of life. Evolutionists want us to believe that the development of the most complex components of life happened during this primordial period immediately after the Earth's environment was ready to support life. However, there is no plausible explanation of how this could have happened. In reality, we have enough evidence to show that life which existed right from the beginning was, in principle, very similar to present organisms.

As a result of this stance we are supposed to believe that life somehow arose and only later developed in some unknown way. Of course there are many hypotheses offering a wide range of scenarios, but none of them are plausible and many are just preposterous. I found that this approach in essence does not differ from the religious approach demanding belief without evidence. So evolutionists have selected a non scientific option in their explanations of the origin of life although using scientific language.

These explanations became unacceptable to a significant part of the scientific community and as a result a new scientific theory of intelligent design has been proposed. This theory postulates that organisms could not have developed as a result of evolution due to the complexity of their biological design, but had to be created by an intelligent designer. Michael Behe in his seminal book *The Darwin's Black Box*[1] presents irrefutable arguments for intelligent design. He also introduced a novel concept of 'irreducible complexity' showing that certain biological components could not arise as a result of random mutations.

This theory provides valid scientific arguments against evolution. Many of these arguments could not be answered by evolutionists, but this does not affect their belief because they associate intelligent design with supernatural phenomena. The problem is that the intelligent design theory does not identify the designer but does not reject God. The weakness of the intelligent design theory lies in the lack of scientific explanation of how and why life arose and developed on Earth and who was responsible for this happening. The purpose of this book is to explain the whole process of genesis and to provide assumptions about the designer.

In spite of so many discoveries in the field of molecular biology, evolutionists still propagate the perception that life is 'simple' and therefore could arise spontaneously. Since academic explanations of the origins of life on Earth were so 'simple', this approach was taken by astrobiologists who assumed that the presence of water and the correct temperature were sufficient for life to arise anywhere. Therefore there must be millions of planets in our Galaxy populated by extraterrestrial beings. This optimistic approach resulted in the scientific program called SETI[2] looking for intelligent life in the Universe. However, more than 50 years of searching has not yielded any results. Where are they? Some scientists started to believe that life on Earth is a very unusual phenomenon and more of a freak than a rule. This was outlined by Ward and Brownlee in their book[3] which described a long list of conditions needed for life to exist.

It appears that there is a need to explain the origin and development of life taking a purely scientific approach based on known facts using scientific methodology, without any bias and pressures which are faced by academic researchers forced to conform to evolutionary paradigms. The general public, which is fed official information originating from the academic establishment, does not realize the hidden disagreements within the evolution camp. It does not realize that the official information is well censored and strongly biased towards supporting evolution, that some explanations are 'one-sided' and are telling 'half-truths'. In reality, the general public is led onto the evolutionary path not realizing that this path is not scientific at all.

One of the main objectives of this book is to tell the truth about the state of evolution using only scientific data. It emphasizes that one should use probability when evaluating the plausibility of evolution. What is missing from the research on evolution is the calculation of the likelihood of key events. We know so much about the structure of key molecular components and the way they are produced that skilled scientists should have enough information to produce some sort of numbers. However, these calculations are missing as one could guess that probabilities would be so low that even 4 billion years would not be long enough to create a living cell.

To illustrate the probability problems of evolution, this book provides descriptions of several very complex molecules to enable readers to decide for themselves if such molecular designs could arise by chance.

The present scientific explanations for the origins of life on our planet uses statements such as "how our planet was extremely lucky to be in the right place at the right time, to have the right sun, to have so much water, the right temperature, tectonic plates, etc, etc." We have to ask how much more luck could one get? If the probability of one lucky event is one in a million and there are half a dozen such lucky events, we are not moving in the scientific domain but in the domain of miracles. If we do not accept miracles we have to find a rational explanation for the causes of these events, and this book tries to provide such explanations.

So we finally arrive at the most important questions. Is life on Earth an improbable phenomenon? Is life in the Universe something unique? If life is such a rare and unusual phenomenon how and why could it arise on Earth? Is it a question of throwing biological dice many times over, or is it the existence of other unknown forces which drives the development of life on Earth?

The main tenet of this book is to show that life on Earth is such a complex phenomenon that it could not have arisen spontaneously, driven by random events and could not have evolved into intelligent organisms. Such life could only have arisen as the result of a purposeful design by intelligent beings. I propose that the arising of intelligent life on Earth was the result of a well planned long term experiment. It is not the outcome of supernatural forces but the product of beings which are very different from us. These beings are not only intelligent but also very powerful as measured by our standards. In a certain sense they could easily be considered as gods. However, regardless of who they are, they have to obey the universal laws of physics.

One of the major objectives of this book is to provide an independent assessment of biological knowledge and evaluate it using methodology used in physics and engineering. I believe that biological systems should be treated in exactly the same way as mechanical or electronic systems, using similar criteria and rules. The fact that we are dealing with biological molecules does not make them different from chemical polymers and other materials used to make the world around us. One could say that biology is a manufacturing enterprise.

This book takes a holistic approach to the subject of life on Earth and the arising of man. Life is such a complex phenomenon depending on so many parameters that all of them should be studied and evaluated. The actual process of genesis is not as described in religious literature. Quite the opposite. It is a very long and protracted process lasting about 4 billion years. It can be divided into several stages as a very logical sequence of interdependent events. It starts with the selection of a suitable planet and ends with the arising of intelligent modern man, the ultimate apex of creation. The most critical event in the whole genesis process is the arising of life itself. There is no direct scientific proof how this occurred as we cannot go back several billion years to check what really happened. But there is a substantial amount of circumstantial evidence showing that life was introduced by an extraterrestrial civilization.

It would be very difficult to prove purposeful design without understanding the basic workings of the cell and its molecular components. However molecular structures of the cell are so immensely complex that at present they are only accessible to scientists specializing in this field. Popular scientific books simply omit descriptions of molecular complexes. Therefore the general public, not having first hand access to this knowledge, has to rely on popular evolutionary explanations. This book tries to fill that gap and to pass to the reader an awareness of the complexity of cells. However, it would be difficult to convey this complexity without describing its basic molecular components and structures. This book tries to keep scientific vocabulary to a minimum, but a certain amount of terminology is unavoidable. These terms are marked with an asterisk and are briefly explained in the glossary at the end of the book.

The first stage of genesis was the preparation of Earth for life. A planet, selected from millions of planets, had to be in the right place in the Milky Way and in our solar system which would guarantee the right environmental conditions. This planet also had to have several unique properties like tectonic plates and a magnetic field. The Earth was selected to be a habitat for life because it met so many special conditions. The process of selection is covered in chapter 1.

The most important ingredient for life on Earth is water. Water has very peculiar properties not shared by other liquids and therefore provides a unique contribution to life. Water was needed in large quantities but did not exist in this part of the solar system. Therefore the delivery of water had to be arranged and this is covered in chapter 2.

After the delivery of water Earth was ready to support first life. The first organisms on Earth were small, bacteria like cells. These cells had the ability to capture the Sun's energy using photosynthesis and were responsible for the generation of oxygen without which higher life would not be able to develop. The first bacteria were already very complex organisms employing several huge molecular structures which laid the foundation for future designs. These structures and their functions are discussed in chapter 3.

For a period of about 2 billion years not much happened on Earth. The oxygen generated by bacteria was changing the planet by oxidizing iron and other metals. Another giant developmental step occurred when a complex cell, eukaryote* arrived on the scene. This type of cell is a basic building block for the edifice of more complex life including man. The complexity of this cell is outlined in chapter 4.

The first cell was in reality a very sophisticated biological system with fully developed energy generation and transformation processes. The energy system is an incredibly complex biological creation surpassing everything man has managed to invent so far. The energy system which was introduced in the first organisms is still used in all organisms on Earth today including man. An overview of this system is provided in chapter 5.

The next 1.5 billion years was a subdued period in the Earth's history. Then around 530 million years ago, during the so called Cambrian explosion which lasted about 25 million years, the major body plans of all living organisms appeared. This was a very important episode in the development of life. This period is covered in chapter 6.

Up until then all life was water based and there was no body design exclusively for terrestrial organisms. Water was the most protective and friendly environment, but in spite of this, life moved onto land, which was an inhospitable habitat. This move was not easy at all and needed significant body changes in animals as well as in plants. This move is discussed in chapter 7.

While the development of the animal world was progressing well, even the most advanced apes did not resemble intelligent beings. There was a need to generate a new species which would have the ability to develop intelligence and consciousness. This could not be a continuation of existing apes because the new brain would have to be completely different to that of all other animals. The development of man is covered in chapter 8.

The apex of the development of life on Earth is without any doubt the human brain. The development and modifications of the brain took place in several stages, the last one occurred about 6000 years ago. The brain is probably the most complex construction that ever existed in the Universe and is discussed in chapter 9.

Public perception and understanding of Darwin's theory is not always correct, and so in Chapter 10 I provide a more truthful explanation of evolution. The two fundamental pillars of evolution, mutations and natural selection are explained and their limitations clarified.

Because evolution was not acceptable to many scientists, a new paradigm of intelligent design was introduced and is explained in chapter 11. Since the intelligent design theory does not include the arising of species, a new theory of directed evolution is proposed.

It is noticeable that biologists treat the natural world differently from the world of physics or engineering. This is probably because organisms look different and are built from organic molecules rather than from metals or plastics. But on closer examination we can see that organisms operate like any man-made machines or factories. The engineering approach to biological structures is presented in chapter 12.

In chapter 13 all the major genesis events are reviewed with the conclusion that they could not have happened by chance and therefore had to be organized by intelligent beings as part of a special 'experiment'. The new genesis scenario is proposed and its important stages are discussed. Identification of the designer responsible for the genesis of life is the key element in this analysis. Whether these intelligent beings are also responsible for human suffering is another important topic discussed in this chapter.

The most controversial hypothesis about the presence of extraterrestrial beings on Earth is discussed in chapter 14. Three unusual events are discussed which should provide plausible proof that in spite of what we believe 'they must be on Earth'.

The picture of genesis that I present is based on the most up to date scientific evidence available. It is important to emphasize that no supernatural forces are involved in this process. The incredible progress in scientific discoveries achieved during the last 20 years has shown that life is a far more complex phenomenon than we could ever have imagined. Only intelligent beings could have been responsible for the genesis of such complex life. There is no direct proof that extraterrestrial civilizations are involved in the development of life on Earth, but looking at the whole process of genesis, this is the most plausible scientific solution.

CHAPTER 1

The selection and preparation of a planet

The prevailing scientific opinion is that there are billions of planets in the Universe and hundreds of millions in our galaxy which could support life. This estimate is based on the very simplistic understanding that it is easy and natural for life to arise over a certain period of time where liquid water is present. This scientific opinion was formulated in 1961 by the so called Drake equation[1] which attempted to predict the number of civilizations in our galaxy with which we could communicate. Drake assumed that life would arise on all habitable planets which would eventually produce intelligent civilizations. He estimated that up to 100 million civilizations could exist in the Milky Way. Even a superficial investigation would show that the number obtained from this equation must be wrong. This is confirmed by the so called Fermi paradox which states that if life in the Universe is so common then why has it not yet been detected?

The environment needed to support intelligent life is extremely complex, but our expectations are still based on the Drake equation which provides unrealistic results. Only the Rare Earth hypothesis[2] looks more deeply into the conditions needed to meet life's requirements. This hypothesis maintains that several specific conditions must exist in order for a planet to harbor life. The conclusion is that there are not that many planets in the Milky Way which could meet these exact requirements.

Life is a very fragile phenomenon which can only exist in a benign environment. Such an environment is described as the habitable zone. It must meet the basic requirements necessary for life such as liquid water, the right temperature and low radiation. However, these conditions are essential but not sufficient to support life. A planet which is going to harbor life must meet many other conditions and these will be covered in this chapter. I will consider the steps which intelligent beings would have had to have taken before selecting the right planet to support their life experiment.

The selection of a planet with the potential for a benign environment must start with an analysis of the conditions in our galaxy and the identification of a suitable zone where life could exist. The next step would be the search for an appropriate sun which would provide a stable energy supply over many billions of years. The position of the planet within the solar system is vital and must also be in the correct zone. The geological structure of the planet, its chemical composition and its atmosphere are also critical parameters for life to survive over billions of years.

The selection of a sun in our galaxy

We are living in the Milky Way which is a spiral galaxy. Spiral galaxies, when viewed from above, look like a disc with branching arms. Our galaxy has a diameter of approximately 100,000 light-years[3] and is about 2,000 light-years thick. The Milky Way has about 400 billion stars, but it is not known how many of them might be suitable to support life.

The central zone of our galaxy is a very active region which has a high density of stars. New stars are born, old stars explode and there are many collisions between stars in this region. It is believed that in the centre of the galaxy there is a massive black hole which is the source of strong radiation known as Hawking radiation[4]. In this region there are many neutron stars and supernovae which are the source of dangerous radiation. The level of high energy ionizing particles, gamma rays and X-rays is very high there making life in this region impossible. It is not known how large this dangerous zone is but it is estimated to extend at least 10,000 light-years from the centre. The protection of life from this high energy radiation can only be secured by separation because the intensity of radiation falls with the square of the distance from the source. Therefore a planet which could shelter life must be as far away as possible from this zone.

However, at the outer regions of the galaxy the concentration of heavy elements necessary for the formation of solid planets is low. To form an Earth-like planet, large quantities of metals, silicon and oxygen are needed which are very rare in the outer regions. Many more elements heavier than helium such as nitrogen, carbon, etc. are also necessary to create life. It appears that the selection of a zone close to the position of our Sun, about 27,000 light-years from the centre of the galaxy, is optimum for offering good safety from radiation and an abundance of planet and life building materials.

There are many stars in this zone but only one of a particular size would be suitable to support life. A suitable star must have strong enough electromagnetic radiation to provide enough energy for an extended habitable zone and must also have a long enough life span to support life for many billions of years.

In the Milky Way the most common stars are red dwarfs which are smaller and cooler than the Sun. But such small stars are not suitable for harboring life. For example, a star having mass equal to about half of that of the Sun would have luminosity of only 3.5 percent of that of the Sun. With such a small energy output a habitable zone around such a star would be very small. Planets in the habitable zone of a red dwarf would be so close to the parent star that they would likely be tidally locked like our Moon is locked to Earth. This would mean that one side of the planet would be in perpetual daylight and the other in eternal darkness resulting in enormous temperature variations from one side of the planet to the other. It is unlikely that this type of star would be suitable to support Earth-like life.

A slightly larger star, say having a mass of 1.3 of our Sun, would have a large habitable zone starting beyond Mars' orbit. However its life span would be less than 4 billion years which would be too short to develop the right planetary conditions to sustain intelligent life for a long period of time. It has also been discovered that ultra violet radiation from stars having a mass between 1.2 and 1.5 of the Sun would be very strong and could cause significant damage to DNA, meaning that larger suns would harm prospective life.

It appears that our Sun perfectly meets all the criteria required of size, position, habitable zone and its lifetime. Our Sun's lifespan is about 10 billion years which should be sufficient to develop and sustain intelligent life. It is believed that only about 3.5 percent of stars are of the G class[5] to which our Sun belongs, but it is not known how many Sun-like stars there are in the Milky Way.

Position of the planet in the solar system

One of the most important conditions to sustain intelligent life is the right temperature of the environment. Since the chemistry of life is based on water, the average surface temperature of the planet should be above the freezing point. Life would find it very difficult to survive over long periods of time in temperatures below freezing. Since cold blooded animals can be killed by temperatures above 40°C, the average temperature should be well below this upper limit. Warm blooded animals can cope much better with extreme temperatures but preferably not much outside the range of -40°C to +50°C. Taking into account the above limitations the average temperature of a habitable planet should be between 5°C and 25°C with extreme temperatures between -40°C and +50°C.

The planet's temperature depends on several factors. The most important factor is how much energy is delivered from the Sun and this in turn depends on the level of the Sun's radiation and on the distance between the Sun and the planet. To receive enough energy the planet must be placed in the right orbit around the Sun. Too close and the temperature would rise above the boiling point[6] of water. Too far and temperatures would drop below zero preventing the growth of life.

The second factor is how much energy is absorbed by the planet and this depends on the planet's albedo, or reflectivity coefficient, which determines how much energy is reflected back into space. The planet's albedo depends on the condition of its surface, for example ice, snow and clouds reflect a large proportion of energy making the planet cooler. On the other hand plants and water absorb a large amount of incident energy. Water vapors forming clouds in the lower atmosphere reflect the Sun's visible radiation increasing the planet's albedo but also prevent the loss of heat by stopping infrared energy radiating into space.

The third factor affecting temperature is the greenhouse effect. The greenhouse effect works by trapping the Sun's energy in the atmosphere. The atmosphere is transparent to the Sun's visible radiation which is converted into heat on the planet's surface. This hot surface radiates back infrared energy. However the infrared radiation is captured by the atmosphere and is sent back to the surface warming the planet. So the system operates in a similar way to a glass greenhouse. Water, carbon dioxide and methane are the three most important greenhouse gases. Recent research shows that water vapor in high atmosphere acts as a strong greenhouse gas by absorbing infrared radiation more effectively than carbon dioxide. Greenhouse gases are eventually broken down by the Sun's radiation and should be replenished to prevent cooling of the planet.

All these parameters affect Earth's climate in a complex way which is not fully understood yet. We know that Mars' average temperature of minus 65°C is cooler than the Earth's would be if placed on Mars' orbit. This is because of the lack of atmosphere and water on Mars. The same applies to the Moon having an average temperature of minus 18°C comparing with Earth's temperature of 16°C. As well as atmospheric composition, the Earth's climate is affected by many other factors such as the presence of large quantities of liquid water, the tilt of the Earth's axis and the resulting changes of the seasons, the speed of the Earth's rotation, daily temperature variations, the orbit around the Sun and the effect of the Moon on the tides.

Planetary orbit, rotation and tilt

The right climatic conditions also depend on planetary orbital eccentricity. Orbital eccentricity is the difference between the planets furthest and nearest approach to its parent star divided by the sum of said distances. It is the ratio describing the shape of the elliptical orbit. The planet's orbital eccentricity should be small enough to prevent large temperature variations occurring during the planet's journey around the Sun. The greater the eccentricity, the larger the temperature fluctuations on the planet's surface as the amount of energy received varies as an inverse square of the planet's distance from the Sun. For example, a planet positioned at double the distance receives only a quarter of the energy.

To secure stable habitable temperatures, the Sun's radiation should be equally distributed over a large part of the planet's surface. Therefore the planet should rotate with a relatively high speed around its axis otherwise surface temperatures during the day and night would fluctuate beyond acceptable limits.

The optimum time for one revolution for an Earth sized planet would be about 24 hours[7]. Seasonal variations of temperature are also important as it secures good circulation of the atmosphere across the planet and water currents in the oceans. This spreads the habitable zone over much larger areas across the planet and reduces extreme temperatures on the equator and the poles.

The tilt[8] of the planet's axis has a very important influence on its climate. It controls the distribution and amount of radiation falling at different latitudes. Good seasonal variations of temperature could be provided by tilting the planet's axis by 15-30 degrees. If there was no tilt, most of the energy would be received on the equator making it very hot whilst keeping the polar region very cold. A larger tilt would cause extreme seasons and a lack of climatic stability exposing life to extreme variations of temperatures. At present the Earth's tilt angle is 23.45 degrees and the polar regions receive about 40 percent of the energy received at the equator. Should the tilt exceed 54 degrees the polar regions would receive more energy than the tropics.

Studying the Earth's climatic history shows us that around 8,000 years ago its tilt moved from around 24.14 degrees to its present value. This small change in the tilt caused large climatic changes. This resulted in a large part of the present Sahara area being transformed into desert. Archaeological discoveries confirmed that before the change, the Sahara was a green, water rich area supporting many plants and animals. The end of the 'Green Sahara' came about quite suddenly, around 5,500 years ago and was caused by the ending of the monsoon which brought water to these latitudes. As a result of these changes, besides the Sahara, other areas such as the Arab peninsula, central Asia near the Aral Sea and Gobi became deserts. Thus a very small change in the tilt led to an abrupt collapse in that ecosystem.

The foundation of a planet able to support life

Size of a planet

For a planet to be able to support intelligent life, it must be large enough to have an appropriate gravity. Small planets having a low gravity would not be able to hold water vapor and gas molecules, which would escape into space. Each planet is characterized by its escape velocity[9], the value of which depends on its mass. For a planet to keep its atmosphere the speed of gas particles in the upper atmosphere must be lower than the escape velocity. For example, using Jeans[10] gas escape calculations, Earth is not able to keep hydrogen and helium in its atmosphere. A planet about a quarter of Earth's mass (0.63 of Earth's radius) and having a temperature similar to Earth would not be able to keep water in its atmosphere. A dense atmosphere is necessary to provide heat and radiation insulation and to support liquid surface water.

Also, a small planet would cool down very quickly and its interior would be too cold and solid to contribute any heat to the planet's energy. Such a planet would not have a magnetic field and tectonic plates needed for the development of life.

A habitable planet must not be too large. A large planet might accommodate bacteria and other small organisms but is not suitable for larger animals. The bone structure of larger animals must be able to support their weight and their muscles must provide sufficient force to accelerate and move them. Normally the bones of animals on Earth can support about ten times the weight of their body, but we have to take into account the dynamic forces acting on the bones which could be very high.

Since muscle and bone strength increases with the square of its body dimensions, and body weight increases with the cube of its dimensions, to operate at higher gravity and to keep the same performance, the body would have to be scaled down. For a body to operate with the same effectiveness at twice Earth's gravity it would have to be a half its original size. Normally the downsizing of a body should not be a problem because there are many hot blooded animals much smaller than man. However we have to consider that man must have a head sufficiently large enough to accommodate his brain, which cannot be reduced. Therefore in a scaled down version of man, this head would be disproportionally large.

In conclusion, intelligent life would not be able to develop on a planet with a gravity of more than twice that of Earth's.

Tectonic plates

In order for a planet to be able to support life over a long period of time it must have plate tectonics. To understand plate tectonics we have to look into the structure of Earth. The cross-section of Earth reveals[11] three concentric layers. Around the outside there is a thin, hard crust ranging from 10 to 100 kilometers thick. Under that is a donut-shaped mantle 2,900 kilometers thick which consists of slow flowing viscous molten rock. At the centre of Earth lies a two-part core. The inner part has a 1,200 km radius and consists of iron and nickel. The outer core surrounding it is an ocean of similar liquid metals 2,300 kilometers thick.

The continental crust is made of low density of rocks such as granites but the ocean bed crust is made of much heavier basalts. Basalt comes from the mantle in liquid form which is visible during volcanic eruptions when it flows as lava. Because granite is less dense than basalt, the continents float on a layer of basalt.

What are tectonic plates? The surface of Earth does not form a uniform solid crust, but consists of several plates which look like a cracked egg shell. Tectonic plates include all the crust and a thin section of the mantle underneath it. Plates composed of the oceanic crust and the mantle are 50-60 km thick, whereas plates with continental crust are about 100 kilometers in thickness. The plates float on the mantle as their density is lower than that of the mantle beneath.

The mechanism for moving tectonic plates can be explained by the following model. Deep beneath the Earth's surface, the hot, fluid upper mantle rises up. When it reaches the top layer it flows parallel to the surface of Earth moving the plates which are floating on top of it. After travelling a certain distance the fluid begins to cool and sink. The oceanic crust made of heavy basalt moves with the fluid and when the fluid starts moving downwards in the subduction zone, it takes with it the basalt plate. Subduction zones are long linear regions where the oceanic crust is driven into the depths of Earth. In such regions, as a result of the interaction of two tectonic plates, linear mountain ranges are formed.

Tectonic plates are important for maintaining the right conditions for life. Active volcanoes and moving tectonic plates sculpture the Earth's surface by forming high mountains and deep oceans. Without moving tectonic plates and volcanoes the surface of the planet would be relatively flat and covered completely with water, so there would be no opportunity for life to develop on land. Whilst volcanoes would form steep mountains, they alone would not be suitable to support complex life. Large continental areas are needed to support intelligent life and this would not be possible without the existence of moving tectonic plates. Without this continuous mountain forming process, existing continents would slowly diminish through erosion and eventually, after 20 to 30 million years, even the world's highest mountains would be reduced to sea level. All this erosion material would end up in the oceans causing water levels to rise, leaving Earth water logged[12]. This erosion is evident now in rivers carrying silt to the sea and waves reducing low level coastlines. The removal of continents would not only eradicate life on land but would also affect life in the seas as ocean life depends on nutrients which are washed down from the land.

The Earth's carbon dioxide thermostat

The Earth's temperature, to a large extent, depends on its atmosphere which is rich in greenhouse gases such as water, carbon dioxide and methane. Greenhouse gases must be constantly replenished as they decompose with time. The half life[13] of carbon dioxide is about a century and the half life of methane is about 8 years.

The primary source of carbon dioxide is the Earth's interior. Normally volcanoes, subduction zones and ocean ridges release vast volumes of carbon dioxide. If the concentration of carbon dioxide in the atmosphere increases, global temperatures rise. However carbon dioxide in the atmosphere is dissolved in rain water forming a weak acid which in turn causes the weathering of rocks and is, in due course, carried to the oceans. Eventually carbon in sea water is absorbed by marine organisms and is stored on the sea floor in layers of limestone. When the ocean floor sinks in the subduction zone, the carbonate rocks decompose at high temperatures releasing carbon dioxide. Without active tectonic plates this carbon dioxide would be permanently locked in the ocean floor or in the interior of Earth never contributing to greenhouse warming. And without this effect the Earth's temperature would be much lower, even changing Earth into a permanent snow ball. Therefore tectonic plates play an important role in stabilizing the Earth's temperature by releasing carbon dioxide from the core. They drive one of the most important recycling processes by returning, to the atmosphere, carbon dioxide locked in carbonate rocks in the ocean floor.

The Earth's magnetic field

The Universe is not a very friendly place for life. In our galaxy there are many sources of high intensity electromagnetic radiation and high energy particles which travel very long distances. Luckily we are a long way from the centre of the Milky Way where most of these sources are concentrated. However our Sun is not a completely benign star. Besides electromagnetic radiation, which ranges from ultraviolet to infrared, the Sun sends out high energy particles like electrons, protons and alpha particles[14] which form a solar wind. These particles are dangerous to life and therefore must not reach the Earth's surface. Earth is protected from these particles by its magnetic field which deflects them. This field, known as the magnetosphere, causes these particles to travel around the planet rather than bombarding the atmosphere or its surface. The magnetosphere is shaped roughly like a hemisphere on the side facing the Sun, then it is drawn out in a long wake on the opposite side. A small number of particles from the solar wind manage to travel to the Earth's upper atmosphere and ionosphere in the aurora zones. The only time the solar wind is observable on Earth is when it is strong enough to produce phenomena such as the northern lights and geomagnetic storms.

Besides affecting biological life, solar winds can strip a planet of its atmosphere by passing its energy to gas particles which in turn increase their speed beyond escape velocity. It is believed that the absence of a magnetic field on Mars is responsible for the lack of atmosphere there. Not every planet has a magnetic field. It is surprising that in our solar system only Jupiter has a stronger magnetic field[15] than Earth, but all other terrestrial planets have none or very weak magnetic fields.

The Sun also emits ultraviolet radiation which can damage exposed life forms. This radiation cannot be stopped by the Earth's magnetic field but luckily for us it can be stopped by the ozone layer[16]. This protecting layer is in the upper atmosphere, about 20 to 30 km above Earth.

Placing Earth in the right orbit

Finding a planet meeting so many criteria and requirements was not an easy task. It is possible that there was a long delay before conditions on the chosen planet stabilized. When Earth was eventually selected as the most suitable planet in our galaxy for the development of intelligent life, it would have been a miracle if the Earth's distance from the Sun, right from the very beginning, was correct. It is most likely that it was necessary to nudge Earth into its required orbit. As already discussed, the habitable zone around the Sun is determined by the Sun's radiated energy and also depends on tectonic plates, the planet's atmosphere and other factors. All these parameters would have to be considered in the calculation of the new orbit. At the time of adjustment some of these factors such as the atmosphere were not present, therefore to calculate the correct orbit their future effects would have to be predicted.

It is known that when the Earth's distance from the Sun is changed by about 10 million kilometers or by about 7 percent, the Earth's average temperature alters by about 8°C (see Appendix 1). Presently, the Earth's average temperature is about 16°C and is optimized for supporting life on Earth. It has been evaluated that the habitable zone in which water remains liquid around the Sun extends from 0.95 AU[17] to 1.69 AU, but the optimum zone for supporting advanced life is much narrower, between 0.95 AU and 1.15 AU giving the average temperature range between 22°C and minus 1°C.

The most important parameter which affects the position of the habitable zone is the Sun's radiation. This radiation is not constant and increases with time as the Sun ages. It is estimated that 4.5 billion years ago, when Earth was formed, the Sun's radiation was about 30 percent weaker than it is now. Therefore selecting the Earth's orbit to secure the position of the planet in the habitable zone, even after several billion years, would have to take into account any future variation of the Sun's energy. As estimated by some scientists Earth is now lying close to the hot limit of the habitable zone, but during the period of Earth formation it was positioned in a much cooler region. There is some geological evidence suggesting that even a few billion years ago whole Earth was covered in snow and ice and the oceans were frozen. Therefore it was necessary to push Earth closer to the Sun.

Adjusting the Earth's orbit could be possible by ramming Earth with another body. A technique suitable for changing the Earth's orbit is presented in Appendix 1. Calculations show that using a small body with a mass less than one hundredth that of Earth would be sufficient to decrease the distance of Earth from the Sun by 10 million kilometers. This in turn could increase the Earth's average temperature by about 8°C which is a very significant amount.

To decrease the Earth's distance from the Sun it is proposed that a projectile was needed to come from the outer solar region, such as the Kuiper belt, which is about 50 AU from the Sun. A head-on collision slowed Earth by about 480 km/sec and then the Sun's gravity pulled Earth onto its new, closer orbit. A second collision was needed to stabilize the Earth's new orbit. It is possible that the colliding body was larger than calculated in Appendix 1.

Although at present we are worried about the greenhouse effect caused by carbon dioxide, the real problem would be if Earth was too far from the Sun and the Earth's temperature was too low. At sub-zero temperatures water in the form of ice and snow would significantly increase the Earth's albedo causing most of the Sun's energy to be reflected resulting in the runaway freezing of the planet. However, increased Sun energy would generate more dense clouds increasing the albedo and as a result decreasing the Earth's temperature working as a temperature regulator.

Let us look into the origins of the collision. Could it be possible that this collision just happened by chance? First, there were not very many large bodies present in the solar system. How would it have been possible for such a body to leave its orbit and come to Earth? This event was very well timed because it happened soon after Earth was accreted but not completely formed.

The more difficult problem was the collision itself. Nowadays the catastrophic danger of asteroids hitting Earth is widely publicized by some fame seeking scientists. But if we look into the history of asteroid collisions it is an extremely rare event. The last impact of an asteroid as small as 10 km in diameter happened about 66 million years ago. The fact is that even when an asteroid is close to Earth there is no guarantee that it is going to hit it. We know that the escape velocity from Earth is about 11.2 km/

sec. This means that a body having a relative speed higher than the escape velocity when passing even very close to the Earth's surface would miss the planet because the Earth's gravity is not strong enough to pull it in. In the last 20 years a few 'near misses' were recorded when small asteroids passed Earth. The closest was in August 1972 when a meteor passed within 57 km of the Earth's surface above the area of Utah in the United States.

It has been observed that a typical speed of asteroids hitting Earth is about 17 - 20 km/sec although much higher speeds have been recorded. Since the Earth's gravity would not be able to pull a body travelling with such a speed, the asteroid would have to be aimed directly at the Earth's surface.

With Earth travelling at a speed of 30 km/sec it would only take her 7 minutes to cross a particular point on her orbit. This means that for a projectile to hit Earth, it must cross the Earth's trajectory specific point within this time. To illustrate the accuracy required, let us consider a very simple example. Let's assume that the body came from the region of the Kuiper belt, about 50 AU away. If the body travelled on an elliptical orbit it would have taken about 64.6 years (Appendix 1). Therefore the timing accuracy must have been about 0.2 parts per million. Such accuracy could not be achieved by chance.

The collision brought two important benefits. It not only adjusted the Earth's orbit but also resulted in the formation of the Moon. The Moon plays a very important role in supporting life on Earth. Its gravitational force causes tidal movements which form two bulges in the oceans. One bulge faces the Moon, whilst the other bulge is on the opposite side of Earth. The difference between the lowest and highest tides can be as much as 16 meters (53 feet). Such large tides cause significant movement of the oceans and coastal waters resulting in fast tidal currents. Nutrients, consisting of biological waste or products of land erosion, are spread by these currents helping biodiversity of life in the oceans.

Special status of Earth

To harbor life Earth had to meet all the above conditions, many of which are listed in Ward and Brownlee's book. Whilst meeting two or three requirements by chance would be possible, the probability of meeting all of them is extremely low, so the authors proposed that Earth was 'extremely' lucky. However the authors did not attempt to risk calculating any probabilities.

To understand if Earth was lucky we could look into our neighboring planets: Mars and Venus. They belong to a group of solid planets built from similar materials, receiving a comparable amount of solar energy and being so close to each other they could be subjected to similar external conditions like asteroid impacts. The information compiled in Table 1-1 illustrated this.

	Earth	Mars	Venus
Diameter (km)	12756	6792	12104
Mass (relative)	1	0.1	0.81
Density (kg/m^3)	5514	3933	5243
Distance from Sun (relative)	1	1.52	0.72
Mean temperature (°C)	15	- 65	464
Atmospheric pressure (bar)	1	0.01	92
Magnetic field	Yes	No	No
Atmospheric composition (%)	N_2 - 78.1 O_2 – 21	CO_2 - 95.2 N_2 – 2.7	CO_2 – 96.5 N_2 - 3.5
Tectonic plates	Yes	No	No

Table 1-1. Properties of terrestrial planets

These three terrestrial planets should be very similar, but Venus and Mars differ significantly from Earth. They do not have magnetic fields and tectonic plates. Why this is the case is not yet fully understood, although many hypotheses have been presented. It is interesting that the atmospheric compositions of Venus and Mars are almost identical, although their atmospheric pressures differ by several orders of magnitude. Although Earth and Venus have similar quantities of carbon dioxide, on Earth it is locked in rocks through the weathering cycle, whilst on Venus it is in the atmosphere. If water was ever present on Venus it would have removed carbon dioxide from the atmosphere and reduced the greenhouse effect, making the temperature much lower than it is at present. If water was ever present on Mars it would not have evaporated or escaped because of Mars' much lower temperature. The fact that out of these three planets only Earth has a large quantity of water proves that water was specially delivered to Earth because it was necessary to support life.

Comments

The selection of a planet that was suitable for habitation was a very complex process. The list of conditions that needed to be met is quite long, but not necessary complete. This list is based on our understanding of life and the functioning of Earth which is still inadequate, therefore it is most likely that more provisions would had to have been considered. The most critical parameter is the distance of Earth from the Sun, and this would have had to be precisely adjusted before life was seeded on the planet.

It has been suggested that Earth was very lucky to meet all these requirements, but such luck belongs in the domain of miracles and cannot be seriously considered. The only viable scientific solution is that intelligent beings were responsible for the preparation of Earth to harbor life.

CHAPTER 2

Water

"Life without water seems unthinkable. Water is the essential elixir of life. Water is truly an intimate structural element of terrestrial biological activity. Molecular water remains more mysterious than atoms that comprise it. Life and water are all but synonyms."[1]

VOGLER

Water is essential for the existence of life and any quest for intelligence in the Universe is based on the search for water. Why does water exist on Earth? Where did it come from? These are fundamental questions that we will try to answer. Water is a substance with amazing properties and living organisms are designed around these features. Therefore it is worth looking at the characteristics of water in more detail.

Molecular properties of water

Water consists of two elements, hydrogen and oxygen. Its chemical formula is H_2O. The hydrogen atoms create a positive electrical charge while the oxygen atom creates a negative one. The water molecule's odd shape with both hydrogen atoms positioned on the same side of the oxygen atom gives water its ability to "stick" to itself very easily and form droplets. Water also sticks very well to other things, which is why it can spread out in a thin film on certain surfaces such as glass.

Water also has a high level of surface tension. Molecules on the surface of water are not surrounded by similar molecules on all sides, so they are pulled only by other molecules deep inside. In comparison to other liquids, it is hard to break the surface of water. Surface tension makes these water droplets round so they have the smallest possible surface area.

In a small diameter tube, water rises because it sticks to the wall and forms a meniscus as a result of surface tension. This effect is called capillary action and takes place in plants when they "suck up" water.

Thermal properties of water

Water has the highest specific heat capacity[2] of all known materials except hydrogen. It takes a lot of heat to raise the temperature of water. This plays a very important role in climate control as oceans take a long time to warm up and cool down effectively working as heat storage tanks. This property also helps to keep the temperature of organisms down when they are exposed to heat.

Water has a very high heat of vaporization[3]. Liquid water will not change easily into steam, preventing lakes and oceans from drying out in hot seasons. The high heat of vaporization also helps to remove excess heat from organisms. Plants and warm blooded animals rely on water vaporization for cooling their bodies.

Water has a high heat of fusion[4]. It can hold a lot of heat energy before it changes into a solid. This property is important for organisms living in water. Even if the surrounding air temperature changes to below freezing, water will shelter organisms from those changes and provides a stable environment.

Below freezing, water changes into ice which has very unusual properties. Normally densities of materials in a solid state are much higher than when in a liquid state, however in the case of water it is the opposite. The density of ice is lower than the density of liquid water and can therefore float on its surface providing heat insulation and protection for living organisms below. If ice was heavier than water it would sink resulting in the whole body of water becoming frozen. Water is also a good heat insulator[5] and its insulating properties are similar to those of bricks.

Importance of water to life

Water played a critical role in the arising and supporting of life on Earth. It affected the Earth's climate, its tectonic plates and was essential for the generation of oxygen. A large part of all living organisms consist of water.

Climate

Water is one of the most important contributors to the Earth's climate and water vapor is one of the most important greenhouse gases. Without the greenhouse effect the average temperature on Earth would be similar to the Moon[6]. Clouds act as a thermostat, significantly affecting the Earth's albedo and the amount of radiation received from the Sun. When the temperature of Earth rises it produces more clouds, which in turn reflect the Sun's energy cooling Earth back down. The oceans act as a massive energy storage tank reducing daily and annual temperature variations. Ocean currents help to distribute heat around Earth by circulating water. Clouds transport water over long distances affecting the climate of land masses far away from the oceans.

Tectonic plates and the circulation of CO_2

Water is needed for the existence of tectonic plates which were discussed in chapter 1. It is believed that without water the Earth's solid crust would be too strong to break making subduction impossible. This is confirmed by the fact that Mars and Venus do not have liquid water nor tectonic plates.

Water is the major contributor to the erosion of land masses. Rain water causes the continuous erosion of land masses, washing important nutrients into the sea which support abundant life on the sea shelf. Water also plays a critical role in the circulation of carbon on Earth as was discussed in chapter 1.

Photosynthesis and oxygen

Water is essential for the production of oxygen on Earth. Planets such as Mars and Venus do not have water nor any significant quantities of oxygen. Oxygen production is the result of photosynthesis in plants, a process which will be described in more detail in chapter 5. In summary, photosynthesis uses energy from the Sun to combine carbon dioxide from the atmosphere with water to form carbohydrates. During this process water is split into hydrogen and oxygen which is released into the atmosphere.

Not everybody is aware that clean water does not absorb much sunlight in the visible spectrum[7]. Low absorption of visible radiation enables organisms, even living in deep water, to carry out a photosynthesis process.

Water in living organisms

Water is the most important component of all plants and animals contributing 50 to 90 percent of their body mass. The body temperature of warm blooded animals is in the range of 36 to 42°C, while cold blooded animals can function from about 3°C to 38°C.

Water is a good solvent for biological macromolecules such as proteins. However the solubility of proteins is independent of temperature in contrast to most other solvents. This property is very important for cold blooded animals and insects whose body compositions are not affected by temperature.

The origins of water on Earth

While hydrogen is the primordial, most abundant element in the Universe, oxygen was produced later, inside the first generation of stars. These stars, at the end of their life, exploded spreading heavy elements into space forming solar nebulas. It was proposed that in such a solar nebula, which was the precursor to our solar system, these two elements were widely present and possibly in the form of water. It is believed that when the solar system was formed, water was present on the giant gas planets such as Jupiter, Uranus, Neptune and Saturn situated in the cold part of the solar system. However, very little water has been discovered on these planets. In fact, more water is present on the moons circling these planets. It is probable that water is more abundant beyond Neptune, in the regions of the Kuiper belt stretching over 20 AU away.

The terrestrial planets closer to the Sun such as Mars, Earth, Venus and Mercury, being in the hot zone, were practically dry and are still dry with the exception of Earth. Only Earth has a huge reservoir of water, more than one billion cubic kilometers. How this happened is still a mystery. What is more interesting is that water arrived on Earth immediately after the right conditions on Earth were established.

Academic hypotheses of the origin of water

There is general agreement that water on Earth had to come from outside the Solar System and was not present on Earth when the planet was formed. There are several hypotheses on how water originated on Earth but there is no consensus on the source of water. Most scientists believe that water came from outside the Solar System in the form of ice-bearing comets, or large asteroids migrating from the asteroid belt. A few researchers think that water is 'home grown' and came from the proto-planetary nebular disk which provided material for our solar system.

Comets have long been considered the leading candidate for the origin of water on the terrestrial planets. This hypothesis was accepted for two reasons. Firstly, it is widely assumed that the inner solar system was too hot for water to be present. Thus an external source of water was needed. Secondly, Earth and other terrestrial planets underwent one or more magma ocean events[8] that some authors believed would effectively remove any existing water.

The solar system formed as a result of the collapse of a large, cold, slowly rotating cloud of gas and dust into a disk that defined the plane of the solar system. The terrestrial planets grew in this accretion disk in which hydrogen, helium and oxygen were the dominating elements. Some of that hydrogen and oxygen combined to make primordial water.

The region of the solar system inside 3 AU was too warm for volatile molecules like water to condense, so the planets that formed there could only form from compounds with high melting points, such as metals (iron, nickel, and aluminium) and rocky silicates. These rocky bodies would then become the terrestrial planets.

Many scientists believe that when Earth was formed it was very hot and dry. They therefore suggest that millions of water-rich asteroids bombarded our planet around 4 billion years ago. Asteroids could come from the asteroid belt which is located roughly between the orbits of Mars and Jupiter, around 2.2 to 3.2 AU from the Sun. It is occupied by numerous irregularly shaped bodies most of which are quite small. At present the total mass of the asteroid belt is approximately 4 percent of that of the Moon, therefore it does not contain much water.

The main drawback of this hypothesis is that a few billion asteroids would be needed, assuming that each one brought about one cubic kilometer of water. The problem remains how such a large quantity of asteroids could have arrived on Earth and at the same time missed Mars and Venus.

The ratio of hydrogen isotopes

One parameter used to help identify the source of water on Earth is the ratio of deuterium[9] to hydrogen, D/H. The ratio for terrestrial water is 1.49×10^{-4}. When this ratio was measured for comets such as Halley and Hale-Bopp, it transpired that it was twice that for the ratio of water on Earth, convincing many scientists that comets could not have significantly contributed to the supply of water on Earth. However the recent data for comet 103P/Hartley 2 shows a D/H ratio that matched terrestrial water's perfectly[10] opening the possibility of comets being a source of water. It is interesting that the D/H ratios in the water of Martian meteorites agree with the values for comets, showing that water on Earth came from a different source than water on Mars.

A team from the University of Michigan's Astronomy department found that the D/H ratio for primordial water is 2.1×10^{-5} which is almost seven times smaller than the ratio for the water on Earth, thus excluding the proto-planetary nebular disk as the source of water.

Based on the D/H ratio measurements the majority of researches agree that water had to come from an external source. It is possible that the water could have originated from asteroids because the D/H ratio in our sea water matches the value found in water-rich asteroids.

Delivery of water

It is becoming more accepted by the academic world that water was delivered to Earth by several huge asteroids or small planets situated in the outer solar system. If we assume that several asteroids were used to deliver water, we have to consider that the collisions of the last asteroids with Earth would cause the ejection of water which had already been delivered during previous impacts. Water carrying asteroids would have to have a diameter of around a thousand kilometers.

Considering several of the most common hypothesis I came to the conclusion that it was most likely only one body which delivered water to Earth. This body could have originated from the Oort Cloud which is a region of frozen objects more than 5,000 AU from the Sun.

To check the feasibility of this hypothesis let's do a few calculations. The body would have to deliver at least 1.4 billion cubic kilometers of water. Assuming that about 20 percent of its weight was water and its density was 2.6 g/cm^3, the diameter of the body would need to be about 1,720 km and the mass of this body would be about one thousandth of the mass of Earth (Appendix 1).

The delivery of water could only be organized and carried out by intelligent beings who selected and moved a suitable body from its orbit and directed it towards Earth. The main problem faced was how to ensure the body would hit Earth. To achieve this it would have been necessary to predict the position of both Earth and the body, and also their speeds at the time of impact. It would be very easy for the body to miss Earth, even passing as close as a couple of hundred kilometers from its surface. So this was a very precise operation.

Appendix 1 shows the calculations of how to bring a body from the Oort Cloud to Earth. A body moving on a circular orbit around the Sun with speed of 297.4 m/sec was slowed down by 293.2 m/sec to change its orbit to the elliptical transfer orbit. Once on the transfer orbit the Sun's gravity would bring it close to Earth's orbit.

A body was nudged from its stable orbit using a propulsion system. To change the orbit of such a large body would have required very powerful engines operating over a very long period of time. It is possible that such a propulsion engine could use energy generated by hydrogen fusion[11]. Since hydrogen would have been abundant on the body it was sufficient to have a suitable reactor which over a period of many thousands of years slowly changed the planet's trajectory and controlled its passage. Calculations show that to change the orbit during 30,000 years of operation a power of 300TW would be needed (Appendix 1). Another possible solution would be to arrange a suitable collision with another body in the same region which could slow down the water carrying body.

During the journey it would be necessary to implement several corrections of the trajectory and speed. Once the body was moving in the right direction, it is possible that gravitational fields of other solar planets were utilized as means of propulsion. We used planetary gravitational fields when sending out exploration spacecrafts such as New Horizons and Cassini Huygens.

The speed of the body when approaching the Earth's orbit was about 42 km/sec, therefore the impact speed with Earth was about 12.2 km/sec. The energy released during the impact would increase the Earth's temperature by a maximum of 159°C if all the kinetic energy was changed into heat. In reality this rise in temperature would be much lower as some energy would be used to eject large amounts of material from Earth. Also at that time the Earth's surface temperature was much lower than at present due to lower Sun radiation and the absence of the greenhouse effect.

In another scenario it is possible that the body which collided with Earth and ejected a large chunk of mantle to make the Moon, as described in chapter 1, was used to bring water as well. However in such a case the mass of the body would have to be larger to provide sufficient impact energy. The collision of such body with Earth was a major event causing partial destruction of the planet. It is possible that a substantial amount of water, which was carried by the body, was lost during the impact but the water which remained was sufficient to form the present oceans. This hypothesis is supported by the discovery that water on the Moon and on Earth came from the same source[12].

Comments

Water is the key component of this life experiment. Its properties are so unusual that one could almost imagine that water was specially invented for this purpose. We can be sure that life was designed around the properties of water.

Earth holds exactly the right quantity of water to have huge oceans and tectonic plates, but not too much to make Earth waterlogged[13]. Such a precise quantity indicates that water was delivered by a planned action. It is most likely that water was delivered by a large body from the remote parts of the solar system which speed was controlled with incredible precision to hit Earth.

Calculations provided in Appendix 1 illustrate the feasibility of such solutions, but in reality the mechanism for the delivery of water could be very different. It is impossible to prove how water was delivered to Earth and where it came from. However, it is certain that water came from elsewhere as a result of a precisely planned operation, and Earth was chosen as its final destination.

CHAPTER 3

First life

Prokaryotes

Once water was delivered and available in sufficiently large quantities, Earth was ready to support life. The next phase was to prepare a design of organisms which would secure the proliferation and continuation of life on Earth. The design should take into account the contemporary (at the beginning of life) and future environmental conditions on Earth. First, organisms must be suitable for the Earth's primeval conditions but at the same time must be able to adapt to changes in these conditions. At that time, about 4 billion years ago, Earth had the right temperature and enough water to sustain life, but it did not have oxygen which would be needed to support more advanced organisms. So the first objectives were to introduce life forms which would prepare the Earth's environment for more complex organisms.

The first life forms which arrived on Earth were single cell organisms. Seemingly simple, their size belies the incredibly complex processes that lie within. The generation and transformation of energy is the most complex of biological systems and will be covered in chapter 5.

The basic unit of life is the cell[1], being able to grow and reproduce in the Earth's normal environment using the available raw materials. Whilst there are bacteria living in hydrothermal volcanic vents feeding on sulphur compounds, such places cannot be classified as normal conditions because these bacteria would not be able to survive anywhere else.

The first cells, called prokaryotes[2], arrived on Earth about 4 billion years ago. The most common prokaryotes are bacteria which have a relatively simple structure and consist of the cell envelope which is composed of the plasma membrane and the cell wall providing structural integrity and protection (Figure 3-1). Inside bacteria there is cytoplasm – a jelly like liquid composed of water, enzymes, nutrients, wastes, and gases in which components such as DNA* and ribosomes* are present. Bacteria use an ingenious propulsion system called flagellum which looks like a whip that protrudes from the cell body. A flagellum is driven by a very complex rotary motor using proton motive force.

In this chapter we will look into some essential functions of the cell and we will try to show the extraordinary complexity of the simplest living organisms. The existing information about bacteria, although still incomplete, is so vast that it would take many volumes to describe it, so in this chapter I will concentrate on the major features of these organisms.

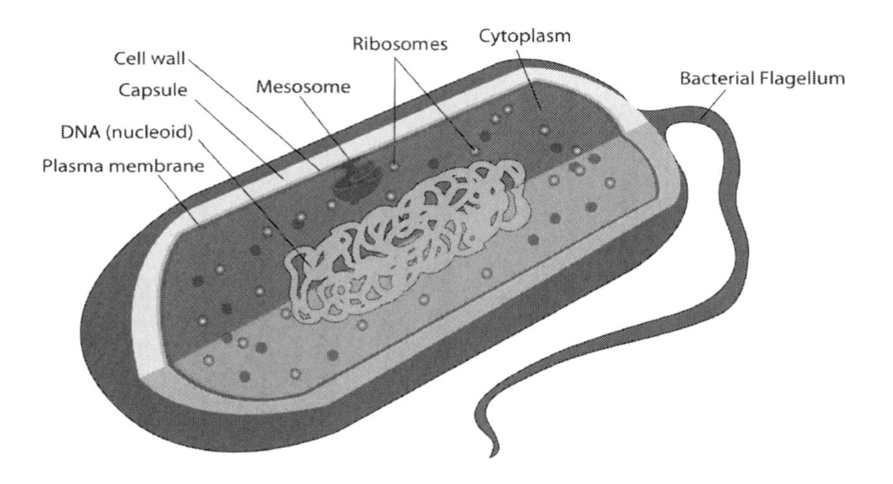

Figure 3-1. Prokaryotic cell – bacterium.

It has been accepted by the scientific world that all life on Earth originated from one common ancestor, but there is no agreement on how this ancestor functioned. So we try to establish what components or building blocks were necessary for this living organism to work. This earliest precursor to life is called LUCA - Last Universal Common Ancestor.

LUCA

The earliest evidence of living organisms on Earth, which was discovered in Greenland, goes back 3.7 billion years which means that life probably started about 4 billion years ago, almost immediately after the Earth's temperature was low enough to sustain life. There have been many hypotheses proposed of the first living organisms. Some evolutionists assume that they could have had a structure much simpler than the structure of present bacteria. They propose simplicity because they want to persuade the general public that this type of structure could arise 'spontaneously' from mineral components existing on early Earth.

I feel that a more realistic and safer approach is to assume that LUCA was very similar to the structure of today's simplest bacteria which has hardly changed for the last 3 billion years. Therefore we can assume that it stayed unchanged right from the beginning of life on Earth.

The first living organisms were not just bags of DNA* with randomly diffused enzymes, but included a sophisticated and complex molecular machinery. There must be a lower limit of an organism's complexity below which biological life is not possible. Genetically modified organisms are now a reality and completely new sets of genes have been implanted in cells, so it should not be too difficult to genetically modify bacteria by removing certain genes and then checking the function of the organism. Such work was performed in the J. Craig Venter Institute where a synthetic genome in *Mycoplasma mycoides*, a parasitic bacterium, was reduced to 473 genes. After the modifications these bacteria reproduced in perfect laboratory conditions at a reduced rate of 3 hours[3]. However the synthetic genome was introduced to living bacteria, therefore the whole supporting cell structure already existed. During this experiment many genes may have been removed which would have affected the long term survival of the bacteria therefore 473 genes might not represent the minimum genome. Since the functions of many genes have not been identified it is difficult to estimate what genome is needed for life.

This organism being parasitic has a metabolism depending on its host and therefore cannot be classified as a typical bacterium. Would this genome be sufficient for LUCA to operate in a real environment? We do not know, but it would be highly unlikely. What should the basic structure of LUCA be which could perform everything necessary for living tasks? It must fulfill several fundamental functions: it must produce and utilize energy, it must store and make use of information on how to build cell structures, it must be able to make all cell components from the available raw materials, and it must reproduce or make copies of itself using the stored information and production facilities.

We know that a very common bacterium like *E. coli** actually has a complex structure and requires more than 4,000 different proteins for growth and reproduction. However LUCA, which must have employed photosynthesis to generate energy, should be similar to cyanobacteria* which have about 3,000 genes. Therefore it is most likely that the first organisms, which would have had to survive in a difficult environment and adapt to varying conditions, would have needed at least 2,000 genes. Some researchers have proposed that LUCA could have had as few as six hundred or a thousand genes but they are not able to provide any substantiation for this hypothesis.

It is not that important how many proteins were used by LUCA, but it is important how they were made. The protein manufacturing process is much more complex than originally thought and it starts with instructions which are coded by DNA*.

DNA

The structure of DNA is itself very intricate because it consists of two strands of biopolymers coiled around each other to form the famous double helix (Figure 3-2). Each strand is composed of nucleotides* including two pyrimidine bases — cytosine (C) and thymine (T), and two purine bases — adenine (A) and guanine (G) which are joined together by the sugar-phosphate backbone. A base pair is created when a purine base in one strand bonds to a pyrimidine base in another strand and vice versa. The two strands are linked by connections between (A) in one strand to (T) in the other strand, and (C) is linked with (G), therefore one strand carries complementary coding information to the other strand. This means that both strands contain coded information. For example, when one strand code is ATTCGCAACCT, the other strand's complementary code is TAAGCGTTGGA.

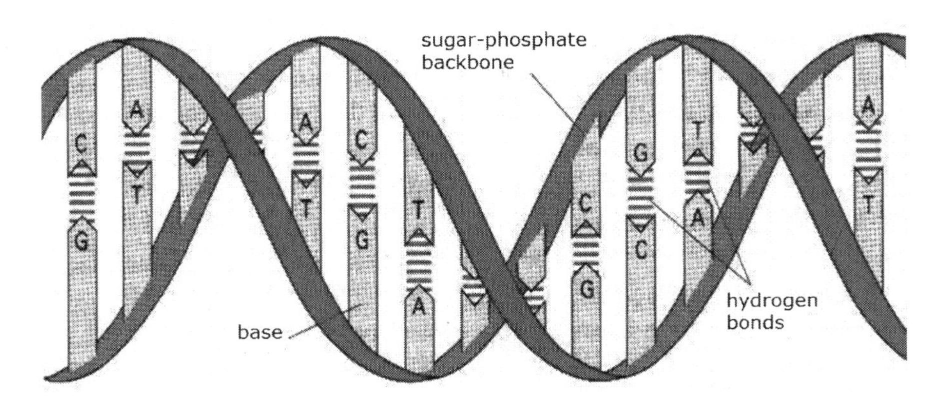

Figure 3-2. Structure of DNA double helix.

DNA stores information about the structure of proteins in a part called the gene. This information is coded by a sequence of A, T, C and G in the strand. One amino acid used in a protein is defined by a sequence of three base pairs of DNA codes which is called a codon. A typical codon could look like TTC or ATC. Only twenty amino acids are used for building proteins although about five hundred different amino acids have been identified. The DNA coding system is the shortest possible as predicted by the communication theory. Three base pairs could code up to $4^3 = 64$ amino acids, but two base pairs could code only $4^2 = 16$ amino acids. In theory probably 16 amino acids would be sufficient to build most of the necessary proteins, therefore a two letter coding system could be used. However, a three letter coding system was selected because it offers a redundancy which helps to reduce coding errors since two or even three different codons can be used to code for the same amino acid. The coding system was selected right at the beginning of life and has not changed since.

Besides protein coding information, DNA also includes additional information which helps to identify the location of the coding sequence. The beginning of the protein code is marked by the start codon and the end of the code is marked by the stop codon. Normally codon ATG signals the start of the protein code and the codons TAA, TAG, TGA indicate the termination of the code.

In the DNA strand there exists another sequence of nucleotides called promoters, which identify particular genes. It is a sort of label which describes the type of protein and it is located at the beginning of the code. So, if we are looking for a specific protein code which could be very long, we have to search only for the right promoter. Promoters could have a length of anything from 100 to about 1,000 nucleotides and are essential for the copying of genes.

Cell division

Reproduction of the cell is a very complex process. There are several steps involved in cell division. Firstly, DNA in the cell must be replicated. Before the replication process starts, the DNA helix must be unwound using a special enzyme called helicase*. Helicase acts as a motor, unzipping the two strands using ATP* as the source of energy. The next stage is to make a copy of one of the strands. This is performed by a special molecular complex called DNA polymerase*. The DNA polymerases are enzymes that create DNA molecules by assembling nucleotides. These enzymes are essential to DNA replication and usually work in pairs to create two identical DNA strands from one original DNA molecule. The process of copying DNA is highly accurate, but a mistake is made for about one in every billion base pairs copied. To achieve such a low error rate the copied DNA strand is "proof-read" by DNA polymerase so that misplaced base pairs can be corrected. This preserves the integrity of the original DNA strand that is passed onto the daughter cells. DNA polymerase is a molecule critical for life and is not affected by mutations[4]. It is very stable and has not changed over billions of years.

A simple cell such as a bacterium normally starts growing by elongation and its internal structures are also replicated. The replication process is very intricate and dozens of special enzymes and hundreds of proteins participate in cell division. The control part of this process is not fully understood yet but what is known shows a very sophisticated regulating system.

Making proteins

Proteins

Of all the molecules found in living organisms, proteins are the most important. They perform a multitude of functions and are essential constituents of the cell's molecular structure. They are used to build bodies, support the skeleton, control processes, digest food and defend against infections.

A protein is made from amino acids which are essential building blocks of the cell. The key elements of amino acids are carbon, hydrogen, oxygen, and nitrogen, although other elements are found in the side chains of certain amino acids. To form a protein, amino acids are joined with each other in the form of a linear chain or necklace of beads. On average a few hundred amino acids[5] join to form one protein, but some muscle proteins can contain up to 30,000 amino acids.

The protein synthesis process (Figure 3-3) takes place in two steps: transcription and translation. In bacteria transcription and translation take place in the cytoplasm and could be coupled, that is translation begins, while the mRNA is still being synthesized. In eukaryotic* cells transcription takes place in the nucleus while translation takes place in cytoplasm.

During the transcription process a part of the DNA which contains the protein coded information has to be copied. This copy is called messenger RNA* (mRNA). RNA is a polymer made from nucleotides like DNA but it has only one strand. This means that RNA does not have a complementary strand. Messenger RNA is made by a large complex molecule called RNA polymerase* which works as a sophisticated and clever copying robot. For example, in the bacteria E. coli, the RNA polymerase molecular mass is about 436 kDa* which corresponds to about 62,000 atoms[6]. This polymerase contains four catalytic subunits and a single regulatory subunit.

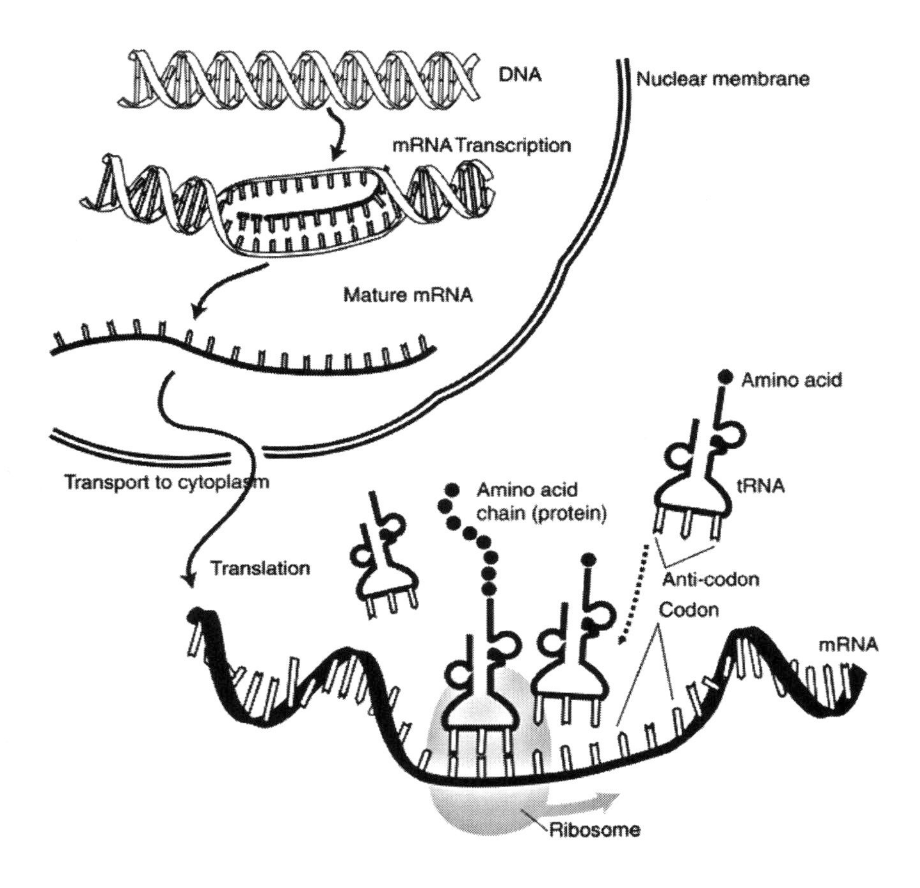

Figure 3-3. Diagram showing the protein synthesis process in eukaryotic cells.

To make mRNA the RNA polymerase molecule goes to the relevant part of DNA and looks for a gene marker called a promoter. Each gene has at its beginning a promoter code corresponding to the specific protein code. However, finding the right promoter is not a simple task and RNA polymerase uses so called sigma factors from the regulatory subunit. There are hundreds of sigma factors and each of them oversees the transcription of a unique set of genes. Sigma factors are discriminatory and each binds a distinct set of promoter sequences.

When the RNA polymerase finds the right promoter it starts splitting the DNA strands. Then the RNA polymerase locks to one strand and starts reading the code. At the same time, it starts producing mRNA, which is a complementary code to the DNA strand code and is therefore identical to the other DNA strand.

The copy of the strand is produced until the RNA polymerase arrives at the stop marker telling it that this is the end of the code. This process operates with remarkably high fidelity. The error rate is less than one mistake for every 100,000 DNA base pairs transcribed. This is due to error correction by an RNA polymerase which "backtracks," or reverses along the transcript to remove wrong or damaged nucleotides. This error rate is much higher than DNA replication as it is not passed to progeny.

RNA polymerase has remarkable properties. Besides automatic error correction which could be compared to the best digital electronic systems, it 'knows' what protein in a given moment is required by the cell. It is still not understood how this occurs.

The next stage of making proteins is called translation where the mRNA produced is then sent to the ribosome – a very large complex of molecules which works as a factory for proteins. To make proteins the ribosome needs amino acids which are floating around the cell. However, how does the ribosome recognize the right amino acid? To do this the amino acid needs a label and this label is provided by another nucleotide sequence that exists in the cell and is called transfer RNA (tRNA)*. Each tRNA has a three letter codon, e.g. AAA for lysine. There are 20 different tRNA units each corresponding to the DNA codes for 20 amino acids. Short strands of tRNA labels lock to the corresponding amino acids as determined by its three letter codon and can then be identified by the ribosome.

But the key question is, how does tRNA with a specific codon recognize the right amino acid? And here the story gets even more complicated. To identify the right amino acid the help of another compound called aminoacyl tRNA synthetase is required. This is an enzyme that attaches the appropriate amino acid onto its tRNA. Each aminoacyl-tRNA synthetase is highly specific for a given amino acid. The chemistry is getting much too complicated and we have to stop here.

Ribosome

The ribosome is a very intricate and elegant cellular machine found in all organisms. It belongs to those indispensable components of the cell which are absolutely essential for life to exist. It converts instructions provided by messenger RNA into the chains of amino-acids called polypeptide chains that make up proteins.

The main constituents of the ribosome are not proteins but non-coding strands of RNA known as ribosomal RNA (rRNA*). Ribosomes are composed of a large and small unit (Figure 3-4), each of which contains its own rRNA molecules and about 50 proteins. The total molecular mass of these two units is approximately 2.66 MDa and they contain about 380,000 atoms. So a ribosome is quite a substantial molecule and many of them are needed in a cell. There is a division of work between these two units. The smaller ribosome unit moves along the messenger RNA and is responsible for reading the code. The larger unit is responsible for joining amino acids and producing proteins.

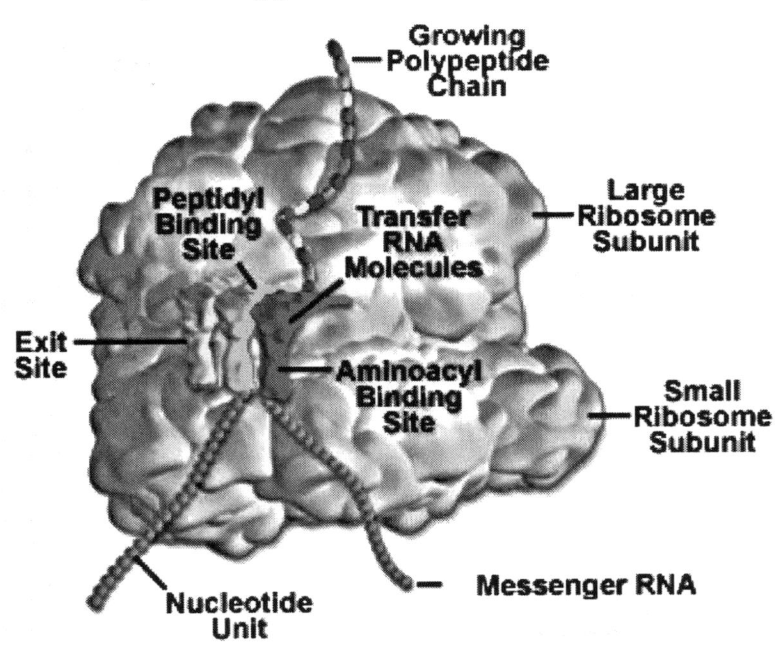

Figure 3-4. Workings and structure of ribosome

The ribosome is vital machinery and is protected against contamination and damage by an external membrane. Molecules entering and exiting the ribosome have to do it through special tunnels in the membrane. Using coding instructions provided by mRNA, the ribosome selects the right tRNA and pulls off one amino acid. The tRNA is then released back into the cell and attaches itself to another amino acid. The ribosome builds a long amino acid chain that will eventually be part of a larger protein. The building rate of proteins is only 10-20 amino acids per second therefore many ribosome units are needed. For example, small bacteria like *E. coli* having 3 million proteins, which every 20 minutes during the cell division must be reproduced, need about 20,000 ribosomes which accounts for about 25 percent of its body mass.

This is a very simplified description of the process of building proteins and highlights only the main stages in their production. Many compounds are involved in DNA translation and protein synthesis and many reactions take place. The process is so complex that to describe it is beyond the scope of this book. This process is common to all living organisms and most likely existed right from the beginning of life on Earth.

Protein folding

After a long chain of amino acids is formed, there comes the next phase of protein construction – folding. In the chain there are amino acids which attract other amino acids, so when the chain folds, bonds are formed between matching amino acids holding the fold together. Folding is a complex process but eventually a three-dimensional protein structure emerges which can take the form of a ball, cylinder or any other shape. This shape must be very precise or else the protein will not perform its function. It is known that a wrongly folded protein, called a prion, can caused brain damage and other diseases. It is a mystery how proteins find the right conformation out of a large number of potential three-dimensional shapes that it could randomly fold into. Special proteins called chaperons have been discovered which help larger proteins fold in the right way. Chaperons also help wrongly folded proteins to unfold and fold again.

When a protein is made, it has to be placed in a specific location in the cell where it is needed. How does the protein find its location? There is another clever system in operation. Protein coding includes specific sequences of amino acids that serve as address labels to direct the molecules to their proper location. How such labels are read is still unknown.

Amino acids

Amino acids are absolutely indispensable for the functioning of life and for cell survival. They not only serve as the building blocks for proteins but also as starting points for the synthesis of many important cellular molecules including vitamins and nucleotides. There are 20 amino acids used by living organisms and their molecular mass is in the range 75 to 250 Da.

It is perceived by the public that amino acids are 'simple' compounds. One is mainly interested in using them as food supplements because nine of them called 'essential' are not produced by the human body. However all twenty amino acids are produced by bacteria and plants. It is perceived that amino acids are widely available in nature and there is nothing interesting one can say about them. However the truth is quite the opposite. The chemistry of making amino acids is as complex as making proteins, if not more so.

Non-essential amino acids are synthesized by relatively simple reactions, whereas the pathways for the formation of essential amino acids are quite complex. Non-essential amino acids can be synthesized from the cell's biochemical compounds in 1 to 5 steps. In contrast, the pathways for essential amino acids require from 5 to 16 steps. However, synthesis of amino acids from raw materials would require many more steps. Six complex enzymes are used in the production of amino acids but the description of these reactions is beyond the scope of this book.

Overview of the protein synthesis process

A summary of the protein synthesis process is illustrated in Figure 3-5. As has already been mentioned the process starts with the transcription of the gene carrying protein coding information by RNA polymerase which generates mRNA. RNA polymerase also generates tRNA labels which are linked to corresponding amino acids. Both mRNA and tRNA with attached amino acids are sent to the ribosome which produces proteins. The process of making proteins is amazingly complex and clever but nothing extraordinary is apparent. However when we look at this diagram showing the interactions and dependencies between different subunits of the protein making system (solid lines) and we add the processes needed to make enzymes used to produce proteins themselves, such as RNA polymerase, ribosome and the enzymes needed for amino acid production, (dashed lines) we are faced with a startling situation. For example, enzymes needed for amino acid production are made by the ribosome which needs amino acids to make them. This means that we cannot make amino acids without already having amino acids. This is the proverbial 'chicken and egg' situation.

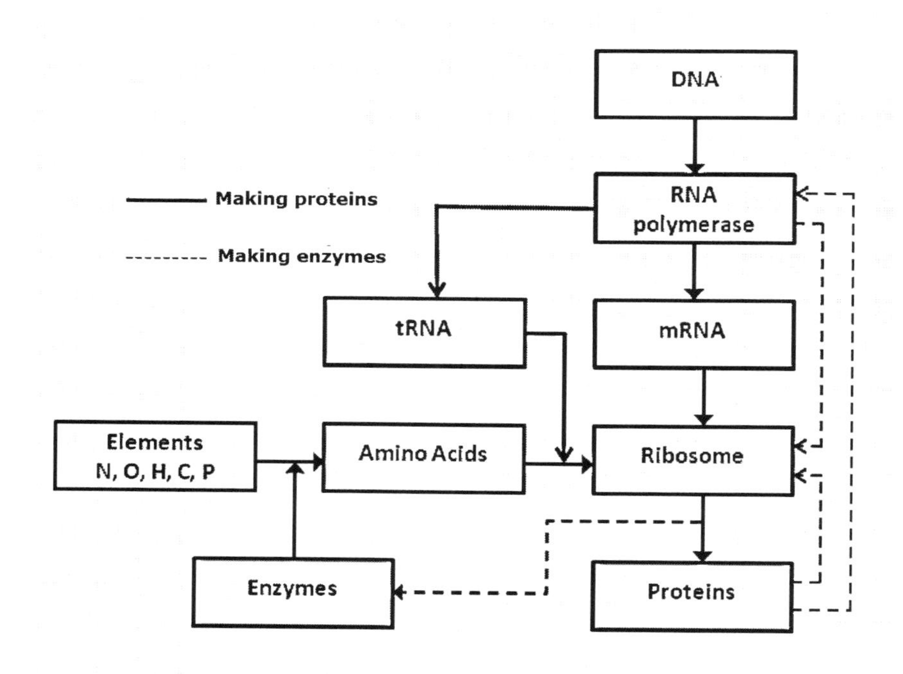

Figure 3-5. Overview diagram of the protein synthesis process.

A similar situation arises when the cell wants to make RNA polymerase for which it needs proteins. However, to make proteins the cell needs RNA polymerase to make messenger RNA and transfer RNA molecules. An even more complex situation exists when the cell needs to make a new ribosome. A ribosome for its construction needs rRNA which is produced by RNA polymerase and also needs proteins which are again produced by ribosomes.

These relationships indicate that to reproduce itself the cell must already have all these necessary manufacturing complexes working. This means that these complexes must have existed in the first cell from the beginning of life on Earth. These complexes are so critical for life that in bacteria they have hardly changed their structure and functioning in over 3 billion years.

Metalloproteins

We are in general familiar with making proteins and although their formation is not straightforward it is well understood. However, we now also know that metals such as iron, manganese, copper, calcium, nickel etc. play a very important role in the functioning of the cell and work as a part of the proteins. What is interesting is that only a very small amount of metal, just a few atoms, added to a protein consisting of many thousands of atoms is sufficient to perform critical catalytic functions. However, the formation of metalloproteins is much more intricate than the formation of proteins and it shows how ingenious and clever the cell's design is. As we know, genes provide information about the coding of proteins by defining the sequence of amino acids, but there is nothing in genes about the formation of metalloproteins, so the cell must find a way around this. Since ribosomes do not receive any information about metals the correct elements must be added to the protein at a later stage.

The biosynthesis of metalloproteins requires two coordinated steps: the acquisition of metal atoms from the outside and delivery to the active site in the cell, and the synthesis of a protein which will link to the metal. The protein must be designed in such a way that besides its normal functions it must have a suitable binding site which will accept a metal ion. Therefore, the acceptance of metal by a protein must be coded in DNA.

However, the acquisition of metals and linking them with the right proteins is not that simple. The problem is that different metals have different affinities* to make a link with a protein. Metal with a strong affinity would kick out a metal with weak affinity which is already linked to the protein. Metals with weak affinities are magnesium, calcium and manganese and metals with strong affinities are nickel, copper and zinc. Normally several of these metals would be present in the cell cytoplasm and they would compete to link to the proteins. Somehow, and this is the clever thing, the cell 'knows' about these affinities and tries to prevent competitive metals linking to the wrong protein. To solve the problem, the cell first puts competing metals in different locations in the cell, then a special protein links to the required metal and folds in such a way that the 'weak' metal is inside the protein and cannot be kicked out by the 'strong' metal when both metals are back together in the same location.

Another clever solution adopted by the cell is to restrict the numbers of metal atoms within the cytoplasm such that 'weak' metals do not compete with other metals for a limited pool of proteins, but rather each protein competes with other proteins for a limited pool of metal atoms. But how does the cell discern the different metals and control their effective intracellular concentrations? The cell membrane has a sophisticated system to control transport of materials in and out of the cell. Special proteins called metal transporters use ATP as an energy source to transfer metals through the membrane. Most importantly, metal transporters have binding sites which bind only the correct metal atoms. The metal transporters make sure that when there is a surplus of metal atoms which cannot find the right proteins they are exported from the cell, otherwise these strong atoms would displace weak atoms during further processing.

Metals imported by the cell are not left floating loose in the cytoplasm but are delivered to the correct protein by special carrier proteins. To help with the insertion of metal atoms into the proteins there are special protein metal chaperons which bind the metal when it is transported into the cell, take them to the right protein and make sure that it is correctly fitted.

To control metal handling by the cell there are metal sensors for each metal used. There are seven groups of sensors controlling about eleven different metals. They make sure that there are just enough metal atoms needed by the proteins. In case there is a shortage of specific metal atoms, the metal sensor sends a signal to the genes to start making more metal transporters. Metal sensors perform many more functions such as storing the surplus of metals in a suitable non activated form. They can control the metabolism of the cell in response to the supply of metals. If the necessary metals are in short supply the cell's metabolism is slowed down until more metal is available.

The handling of metals by the cell is achieved by a very well designed system. It starts with DNA which codes not only metal specific proteins but also metal specific transporters, chaperons, carrier proteins and sensors. This system operates as a sophisticated mechanism controlling the import and export of metals, initiating the building of transporters and influencing cell metabolism. It is part of a huge cell control system which gives us a glimpse of the philosophy of cell design. There is no need to emphasize that this system can only operate as a whole unit and could not have evolved in separate parts via small improvements.

It is estimated that about 30 percent of all types of proteins are metalloproteins. They play a critical part in cell metabolism. In my opinion the two most important cell functions which are controlled by metalloproteins are the splitting of water into oxygen and hydrogen and the fixing of atmospheric nitrogen. The water splitting enzyme will be described in chapter 5 and nitrogen fixing is described here in more detail.

Nitrogen fixation

The growth of all organisms depends on the availability of mineral nutrients, and none is more important than nitrogen, which is required in large amounts as an essential component of amino acids and nucleic acids. There is an abundant supply of nitrogen in the Earth's atmosphere in the form of the molecule N_2.

However, the N_2 molecule is unavailable for use by most organisms because there is a strong triple bond between two nitrogen atoms, making the molecule almost inert. In order for nitrogen to be used for growth it must be changed into ammonia (NH_3) or nitrogen dioxide (NO_2). This process is called nitrogen fixation and requires a catalyst and a lot of energy to achieve it. In industries producing fertilizers the fixation of nitrogen is typically carried out by mixing it with hydrogen gas over an iron catalyst at about 500°C and at a pressure of 300 atmospheres.

However, the process in nitrogen fixing bacteria is carried out at room temperature and atmospheric pressures using only a catalyst. So the catalyst is the clever component. An enzyme called nitrogenase is responsible for this catalytic function. It consists of two proteins: an iron containing protein made up of approximately 9,000 atoms and a molybdenum-iron containing protein of about 31,000 atoms. At the heart of the catalytic process is FeMoco – a cluster of seven iron, one molybdenum and nine sulphur atoms (Figure 3-6.) The biosynthesis of FeMoco is a very complex process and involves nine special proteins and a description of this process is beyond the scope of this book.

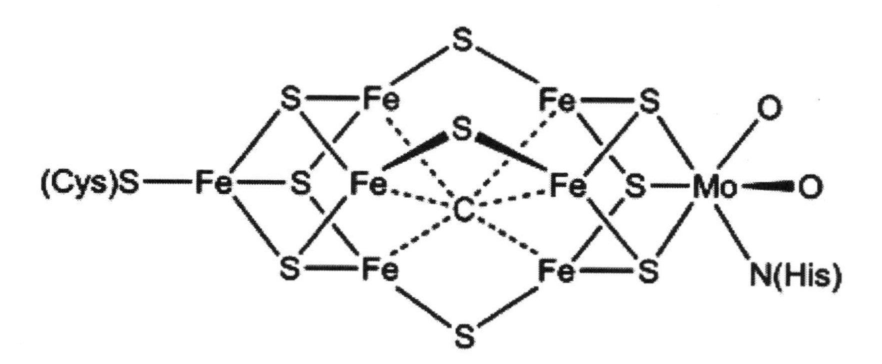

Figure 3-6. Structure of FeMoco – a nitrogen catalyst.

The working of this catalyst is a mind boggling process and is still not fully understood. This process uses a lot of energy and to make two molecules of ammonia, 16 molecules of ATP are needed. This is a good example of how ingenuous molecular designs are, compared with the brutal force often used in man-made solutions.

It is difficult to envisage how evolution could invent such a complex system. How evolution could select such a special metal cluster interacting with dozens of amino acids. Again, DNA codes not only this huge catalyst molecule, but also nine auxiliary proteins helping to assemble the metal cluster. How could DNA know in advance what to code? Besides these molecules the DNA also has to code metal transporters, sensors and many other proteins involved in this highly precise process.

The cell control system

Recent research shows that the internal workings of a cell are not a chaotic, simple process, but precise and complex in which each molecule has its own assigned place and defined functions to perform[7]. Biological components coded by the genes are not randomly distributed in the cell but their position is determined by the cell's controlled system.

In the last decade a team at Stanford University School of Medicine under the leadership of Prof Harley McAdams, who is himself a physicist with extensive work experience in electronics and control systems in industry, published significant discoveries about the cell's control system[8]. His team was working on genetic regulatory circuits of a simple bacterium called *Caulobacter* which lives in clear water. In his opinion these types of circuits do not differ from electronic control circuits.

He discovered that the control circuitry that directs and paces *Caulobacter* cell cycle performance involves the entire cell operating as an integrated system. The cell cycle system is comprised of multiple modular subsystems that implement cellular growth and reproduction. An integral control system constructed using biochemical and genetic logic circuitry organizes the timing of initiation of each of these modular functions. This control circuitry monitors the environment and the internal state of the cell, including the cell structure, as it organizes activation of cell cycle subsystems and *Caulobacter* cell division.

Now we know that the *Caulobacter* cell cycle control system has been exquisitely optimized as a total system for robust operation, coping with variations in metabolic reaction rates and nutrient supply. It reliably stops and restarts the reproduction cycle to adjust to nutrient availability.

The control system utilizes various sensors which monitor the environment and cause the cell cycle progression to slow or stop depending on the nutrient levels, oxygen monitors or internal state monitors e.g., DNA damage.

The *Caulobacter* cyclical genetic circuit is comprised of the cyclically varying concentrations of the four master regulatory proteins that directly control the timing of transcription of over 200 genes. These proteins are synthesized and cleared from the cell one after the other over the course of the cell cycle. Each process activated by the proteins of the cell cycle engine involves a cascade of many reactions. The longest cascade is DNA replication, which involves about 2 million DNA synthesis reactions over about 40 minutes.

Prof McAdams' main conclusions are that "the system design is found to be eminently rational and indeed consistent with good design practices for human designed asynchronous control systems". The control system is a highly organized machine that is rigorously controlled by relatively simple biochemically based control logic, in comparison to human engineered electronic devices.

Analysis of the *Caulobacter* system provides further evidence that biological regulatory systems conform to principles that engineers use to design regulatory systems in other engineering domains. The cell cycle control circuit includes numerous features to ensure that the cell cycle completes successfully under all contingencies. And he concludes that "all these observations and others not addressed here support the observation that it takes a cell to make a cell".

Oxygen

We might not be aware that without oxygen the development of higher organisms would not be possible. We know that some bacteria can manage without oxygen using special chemical compounds to generate energy, but these processes are very inefficient and can generate only a small amount of energy. Multi cellular life needs a large quantity of energy and animals can only obtain it by burning food using oxygen during the respiration process.

Before the emergence of photosynthesis, a process which produces oxygen and continues to replenish our supply, Earth's atmosphere had no significant quantity of oxygen whatsoever. Every oxygen atom in the air was once part of a water molecule, released by photosynthesis. During photosynthesis, a very clever process called water splitting takes place, which will be described in chapter 5. It is this process which is critical for the supply of oxygen on Earth.

About 3.5 billion years ago, a new kind of life was established, which was capable of using the Sun's energy to convert carbon dioxide and water into food with oxygen gas as a waste product. These organisms were photosynthetic microbes called cyanobacteria. They lived in shallow seas, protected from full exposure to the Sun's harmful UV radiation. They belong to that group of organisms which changed the Earth's environment. Cyanobacteria became so abundant that about 2.6 billion years ago the free oxygen they produced began to accumulate in the atmosphere. By about 2 billion years ago the oxygen level increased to about 18 percent of present oxygen levels or 4 percent of the atmospheric level.

The surface of the Earth began to change when the released oxygen started to oxidize iron in the rocks and water. Iron was changed into iron oxides which provide such a characteristic red color. At present all available iron ores are in the form of oxides.

The effect of oxygen was profound. High in the atmosphere, oxygen formed a shielding layer of ozone, O_3 which screened out damaging ultraviolet radiation from the Sun and made the Earth's surface habitable. Nearer the ground, the presence of breathable oxygen opened the door to the existence of whole new forms of life. This oxygen level could support more complex life and around this time eukaryotic cells were developed.

About 750 million years ago green algae came into existence which were more suitable to living in shallow waters and utilizing strong light. It is believed that oxygen levels increased to the present level about 600 million years ago enabling the Cambrian explosion, when most of the progenitors of present animals came into existence. It is considered that in subsequent times the atmospheric content of oxygen fluctuated between 10 and 35 percent peaking during the Carboniferous period lasting from 359 to 259 million years ago.

Comments

Looking at the structure and operation of the first cells we could come to the conclusion that this type of cell is not simple at all. It is called "simple" because it is less complex than the eukaryotic cells which arrived 2 billion years ago. However, its energy generation and respiration processes are as complex as they are in higher organisms. The protein manufacturing process, although slightly different than in the more advanced eukaryotic cells, is still highly complex. To reproduce itself, this cell must have had all these necessary manufacturing complexes working from the very beginning of life on Earth. Therefore it is not possible that they could have developed gradually over a long period of time.

The building of metalloproteins is another example of the working complexity of the "simple" cell. The handling of metals by the cell is carried out by a very well designed system which includes metal binding proteins, transporters, chaperons, carrier proteins and sensors. It is part of a very sophisticated cell control system which could not have evolved in small steps as it can only work as a whole system. Operation of the cell control system is not defined by DNA[9], therefore it must be a part of the cell structure introduced at the beginning of life.

The simplicity of the first cells is pervaded by evolutionists because it is easier to accept the evolutionary theory if the first organisms were very simple. However, the opposite is in fact true. Operation of the simple cell on a molecular level is almost as complex as that of cells developed later and its energy generation and utilization is practically identical to complex cells. In spite of many years of work on the origins of life, there is no plausible hypothesis of how these first organisms came into being.

CHAPTER 4

The complex cell

Arrival of the complex cell

The purpose of introducing photosynthetic bacteria was to prepare Earth for the arrival of more complex life. It took about 2 billion years to change the Earth's environment to be ready for the next stage of life. During this time there occurred a noticeable increase in the level of oxygen, up to about 18 percent of the present level. However, throughout this period the only form of life was still a prokaryotic cell which had not changed much. As a matter of fact bacteria, while adjusting to different living conditions, have survived in the same form to present times. It appears that no significant development of life took place.

Suddenly, about 2 billion years ago[1] life started changing. On Earth, or rather in the water, new and more complex organisms appeared called eukaryotic* cells. The eukaryotic cell is very different from bacteria because it has a very intricate structure and includes many new constituents such as a nucleus*, mitochondria*, chloroplasts* and other organelles. The eukaryotic cell is not only more complex, bigger and has more genes and larger DNA than bacteria, but its processes are much more intricate than that of prokaryotes.

The main feature of the eukaryotic cell is its ability to form higher, multicellular organisms. The prokaryotic cells, while living together in colonies, were never able to link together to form a different organism. But the new cell enabled the construction of higher organisms such as plants and animals. It became the building block of all higher forms of life including man. What is amazing is that the human brain, liver, skin, heart, kidneys, etc. are all built from the same type of cell which is adapted to perform various tasks. But because all the cells are the same, in the body they can cooperate with each other as one entity. So the arrival of the eukaryotic cell was a revolutionary step in the development of life.

Evolutionists recognize that the appearance of the eukaryotic cell was a ground-breaking step because it has been agreed by most scientists that it occurred only once in the history of life. However even evolutionists do not suggest that it came into existence as a result of mutations and natural selection because it is so radically different from preceding organisms.

It was proposed by Prof Lynn Margulis that the eukaryotic cell came into existence as a result of symbioses between bacteria. In her opinion mitochondria were bacteria which were swallowed by another type of bacteria but were not digested, instead they were welcomed as equal partners and started living inside the parent bacteria. This was a very smart solution to overcome the credibility of the evolutionary origin of eukaryotes. Evolutionists unwillingly accepted this hypothesis because they were unable to see how mutations and natural selection could have produced such an incredibly large jump in cell complexity.

Compared with prokaryotes, the eukaryotic cell (Figure 4-1) has a much more complex structure. It has many special organelles which are devoted to particular tasks in the cell. The most important organelles such as mitochondria and chloroplasts will be discussed in chapter 5. Besides the nucleus and mitochondria, the main structures include plasma membranes*, sealed vesicles*, Golgi apparatus* and cytoskeleton*.

Let us look in more detail at the organization and functioning of the cell. The construction of the cell is very complex because it contains several organelles made from thousands to several millions of molecules. For example a small cell of about ten microns can have as many as 50 million molecules. These molecules form very intricate structures and perform very advanced functions. I will try to highlight some of the most important cell components to show how sophisticated and complex their functions are. To cover the functioning of the whole cell is beyond the scope of this book.

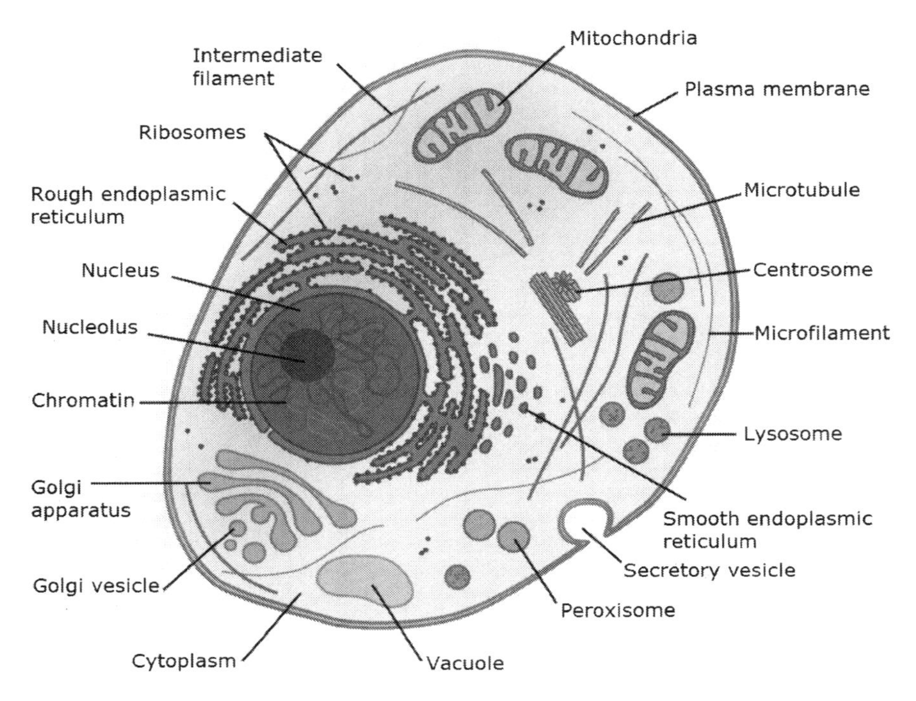

Figure 4-1. Outline of the eukaryotic cell.

Nucleus

The cell's nucleus contains DNA and performs several critical functions. There are two specialized structures inside the nucleus which are factories for making large specialized molecules. One structure, DNA polymerase, is involved in the replication of DNA strands during the reproduction process. Another structure, RNA polymerase is engaged in the synthesis of RNA. All types of RNA such as messenger RNA, transfer RNA and ribosomal RNA which is used to make ribosomes are made in the nucleus.

The nucleus is protected by inner and outer nuclear membranes which form an impermeable barrier. The outer membrane faces the cytoplasm* and the inner membrane faces the chromosomes. Movement into and out of the nucleus occurs through pores where the inner and outer membranes are fused together. Transport through the membrane is controlled by nuclear pore complexes, each consisting of about a thousand proteins. Each pore complex is large enough to accommodate the passage of ribosomal subunits, which exit the nucleus after being assembled in the nucleolus. Materials needed to build ribosomes are delivered through the pores.

In the nucleus there is a DNA supporting structure which works as scaffolding. The gene replication machinery is attached to the scaffolding that allows organized duplication of DNA. For example, in mammalian cells, the nucleus arranges orderly packing of about 0.7 meters of DNA strands inside a sphere of approximately 5 microns in diameter. Remarkably, this structure is completely disassembled when cells undergo replication.

DNA

DNA in the nucleus is arranged in straight chromosomes, often in pairs. The genes are wrapped in proteins which provide protection against accidental damage. These proteins also play a very important role in gene expression[2]. This means that they affect which gene is activated and used to produce proteins. The genome in eukaryotic cells is much bigger than in bacteria because besides a larger number of genes it also has more so called 'junk' DNA which does not code proteins. In humans only 2 percent of the genome codes proteins, but this does not mean that the rest is useless, it's simply that we do not know much about it.

DNA protein coding sequences in the eukaryotic genes are not continuous, as in bacteria genes, but they are separated by non-coding sequences called introns*. Each gene may even have several introns in the sequence. These introns are copied by RNA polymerase and are present in the precursor messenger RNA. However, such mRNA cannot be sent to the ribosome because it would cause an error in the protein chain. Therefore, the introns must be removed from the mRNA and the coding pieces must be spliced together. This work is done by a huge and intricate molecular machine called a spliceosome. The spliceosome is a molecular complex comprised of five small nuclear ribonucleic proteins* (snRNP) and over 300 proteins consisting of more than 700,000 atoms. The picture is complicated because the spliceosome is a dynamic complex and not only changing its form and function during the operation but some snRNP units also perform multiple functions. The whole process of splicing is still full of mystery and it is believed that the spliceosome is the most complicated RNA-protein complex inside the eukaryotic cell.

The purpose of introns is still not fully understood, and some evolutionists treat them as genetic parasites. However, the situation is not that simple because introns take part in alternative splicing. Alternative splicing is a process that results in a single gene coding for multiple proteins. In this process, particular coding parts of a gene may be included within or excluded from the final processed messenger RNA (mRNA) produced from that gene. This means that different, non sequential coding parts of the gene can be spliced together producing a new protein. As a result of alternative splicing the human genome generates 4 times the number of proteins than the number of genes.

The human genome has about 25,000 genes and on average each gene has about 9 separate coding parts and about 12 introns. An introns length varies over a huge range from about 20, to more than 100,000 base pairs, with an average length of about 3,500 base pairs. We know that introns must be removed for the proteins sake but we do not know what they are for. It has been shown that introns affect gene expression, meaning they have influence over which product is made by the genes.

Another interesting element of DNA which recently caught public interest is a telomere. Telomeres are the caps at the end of each linear strand of DNA that protect chromosomes, like the plastic tips at the end of shoelaces. Telomeres get shorter each time the cell copies itself, but the important section of DNA stays intact. Eventually telomeres get too short to do their job, causing the cells to age and stop functioning properly, hence acting as an aging clock in every cell.

Cell membrane

The cell is enclosed by a membrane (Figure 4-2) which performs many different functions. The cell membrane is one of the most important cell structures and not only protects the cell against external mechanical and chemical elements and aggressive organisms, but it regulates the transport of selected chemical components to and from the cell. Membranes are composed of lipids*, proteins and carbohydrates. The lipids are arranged in a bilayer, with hydrophilic* heads facing outwards and their hydrophobic* fatty acid tails facing each other in the middle of the bilayer.

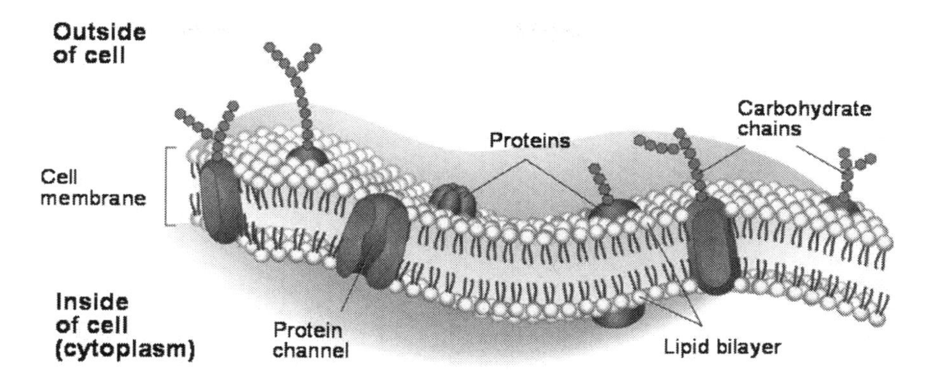

Figure 4-2. Cell membrane structure.

This hydrophobic layer acts as a barrier to all but the smallest molecules, effectively isolating the two sides of the membrane. The membrane does not have a homogeneous structure, but in it various proteins are inserted. The proteins usually span from one side of the lipid bilayer to the other. Various proteins comprise about 50 percent of the mass of the membranes, and are responsible for most of the membrane's properties.

The membrane operates as a very complex, precise and intelligent filter which knows what should be allowed to pass through it and when. It can allow a particular substance to pass through at a certain time, and then reject the same substance at other times. A typical membrane can contain thousands of different types of filters which control the passage of ions, elements, different sized molecules etc.

There are five main methods by which substances can move across the cell membrane: lipid diffusion, osmosis, passive transport, active transport and endocytosis. The first three are passive methods and the last two are active. Passive transport takes place through different sized canals with the help of special proteins. Active transport utilizes several unique types of pumps which need energy to function.

In lipid diffusion, substances can diffuse directly through the lipid bilayer part of the membrane. The only substances that can do this are lipid-soluble molecules or very small molecules, such as water, oxygen and carbon dioxide.

Osmosis is the diffusion of water across a membrane. It is in fact just normal lipid diffusion, but since water is so important and so abundant in cells it is listed separately.

Passive transport is the transport of substances across a membrane by trans-membrane protein molecules. Embedded within the membrane are a variety of transport proteins which act as channels that move molecules into and out of the cell. The transport proteins tend to be specific for one type of molecule so a substance can only cross the membrane if it contains the appropriate transport protein.

Active transport is the pumping of substances across a membrane by a trans-membrane protein pump molecule. These proteins are highly specific, so there is a different protein pump for each molecule to be transported. Pumping is an active process, and it is the only transport mechanism that can transfer substances from lower concentration regions to higher concentration regions.

Endocytosis is used for the transfer of large molecules into a cell. Materials to be brought in are enclosed by an internal fold of the cell membrane, which then pinches its walls to form a closed bag. This bag is then moved inside the cell where its contents are consumed.

Some writers call the cell membrane the brain of the cell because it also reacts to chemical, thermal and mechanical external stimuli. It contains transducers which receive signals from the environment and pass them to the cell control system. There are transducers sensitive to practically all environmental signals such as temperature and light, and to thousands of different chemical compounds such as water, salt, enzymes, proteins, sodium, calcium, potassium etc.

As we learn more about the cell membrane we discover an amazing variety of its different functions. So the membrane, which was originally assumed to only protect the insides of the cell, has become a very complex and critical element which effectively controls the functioning of the cell.

Cytoskeleton

Every cell needs a structure to support its internal parts and aid in its functioning. *Such an objective is fulfilled by a cytoskeleton. It is a sort of scaffolding which is* flexible but stable and it consists of different types of supporting structures such as microtubules, microfilaments and intermediate filaments linked in a complex construction. Microtubules maintain cell shape, provide mechanical support and fulfill many other functions. They are hollow tubes of about 24 nanometers in diameter. Their length varies from a fraction of a micrometer to a fraction of a millimeter.

The transport of molecules in the cell is performed by special motor proteins which use ATP as their energy source. Microtubules facilitate intracellular transport acting as a monorail for vesicle and organelle movement by motor proteins. In most cells microtubules are organized in a radial array extending from a single site generally positioned near the nucleus. This organization produces a network of microtubule tracks along which membrane-bound vesicles are moved. Transported vesicles include organelles such as mitochondria, as well as secretory waste.

The cytoskeleton is extremely dynamic, meaning that the filament systems are able to lengthen or shorten very rapidly which is necessary for the cell to be able to change its shape and complete division. The cytoskeleton also helps the cell move in its environment. The rapid length change of microtubules allows cells to facilitate the reorganization of tracks important for the delivery of vesicles to sites throughout the cell and to build other structures.

Microfilaments are fibers about 7 nm in diameter which play an important role in muscle contraction, cell division, cell movement, and other cellular functions and structures. The filaments provide structural reinforcement, anchor organelles and enzymes to specific regions of the cell and keep the nucleus in place.

The cytoskeleton also takes part in controlling metabolism, sending signals and directing genetic programmers. As we can see it fulfils very important and complex functions, and yet we still do not know how it performs them.

What is coded by the genes?

For more than 50 years scientists have been convinced that all the information needed for the proliferation of life on Earth is coded in the DNA and that genes contain all the data necessary to build a cell – the fundamental block of all living organisms. However, this might not be the case and the latest scientific discoveries show that our understanding of cells should be revised.

In 2005 Franklin M. Harold, Professor of Biochemistry, Biophysics and Genetics at the University of Colorado Medical Centre published a review paper providing the current state of knowledge about the structure of the cell. He stated:

> "The spatial organization of cells, including the arrangement of cytoplasmic constituents and the cells' global form, is not explicitly spelled out in the genome. Genes specify only the primary sequences of macromolecules, portions of which are indeed relevant to the localization of those molecules in space. But cell architecture, for the most part, arises epigenetically from the interactions of numerous gene products. Many of these interactions can be well described as instances of molecular self-organization, either self-assembly or dynamic self-construction. However, when self-organizing chemistry takes place in a living cell, it comes under guidance and constraint from the cellular system as a whole." [3]

So genes provide information on how to build basic blocks of life such as proteins, enzymes, grand molecular complexes, etc, but genomes apparently contain no genes that specify cellular forms such as a membrane or cytoskeleton. So genes specify molecular parts, but not their arrangement into a higher order. In conclusion, the architecture and the functioning of the cell is not determined by its DNA.

How does the cell know how to reproduce itself? It simply copies existing structures and for example, the major classes of cellular membranes grow by extension of the existing membranes. As the cell grows and divides it models itself upon itself. Newly made proteins and other molecules are released into the body that already has spatial structure, and this framework ensures that the placement of new molecules is harmonious with the old structure. Many cellular features and functions such as growth and division are fundamentally dependent on processes that have a specific location in the cell's space. Moreover, these spatial parameters are defined by mechanisms that are, at least in part, independent of genes.

It is not known how molecules coded by DNA make complex constituents of the cell, and what determines their position and operation inside the cell. We do not know the details of how large cell structures such as ribosomes came into existence and what controls the functioning of the cell. However, we do know that the breaking up of a cell into its basic molecular constituents does indeed destroy the spatial organization that makes that cell alive. It has been proven beyond any doubt that DNA is not responsible for the architecture of the cell but the cell itself.

Coding information for the structures of the cell biochemical components, mainly proteins, is preserved in the genes and their copies are passed to the cells. The DNA strands are subject to mutations and therefore the cell building molecules can change. However, cell architecture does not change because it is not affected by mutations. Many parts such as new membranes, cytoskeletons or mitochondria arise as a result of growth of the old structures and therefore they are their exact copies. These structures are completely independent of the genetic processes and therefore are not subject to random detrimental changes. This is why the cell's functions and basic structures have hardly changed over 2 billion years.

What is interesting is that cells can acquire new characteristics as a result of interaction with the environment, and these characteristics are inherited by the following generations. For example, occasionally the daughters of a dividing cell fail to separate and fuse back to back producing a doublet cell. The astonishing finding is that from then on, this one doublet cell propagates indefinitely, subsequent generations of doublet cells and it has been shown that no change to DNA takes place in them.

How does academia deal with this problem? Evolutionists try to explain the origins of these structures by a mechanism called 'self-organization'. Meaning that these incredibly complex structures, some including millions of different atoms, came in to existence by themselves by the accidental arrangement of molecules in the right order. To prove the existence of self-organization in nature, evolutionists cite crystals, flocks of birds or termite nests. It is difficult to see any connection between the growth of a crystal which is made from the same kind of atom or molecule and for example, the growth of complex membranes made from thousands of different components.

The self-organization hypothesis is based on some experiments which showed that certain polarized protein molecules in a test tube can join together and form a sheet or a tube. But these self organized structures are in general homogenous and are formed in a solution consisting of a large number of one type of protein. These structures may remind us of crystals because they tend to have unpredictable dimensions. However, in the cell there is no unlimited supply of molecules, but the cell produces just enough proteins to build the necessary structure. These proteins are transported to the right place at the right time and form a structure having precisely determined dimensions. It is difficult to envisage how such a complex structure like a membrane with many thousands of different built in proteins could arise as a result of self-organization. Since we are told that evolutionary changes originate in the DNA, if there is no information in the genes then how would the cell know that certain proteins would join together and form the required structure and how would it pass this information on to subsequent generations?

Comments

The arrival of the eukaryotic cell is shrouded in great mystery. Evolutionists have had to admit that it could not have arisen as a result of evolution. They therefore invented a new, non-evolutionary mechanism called symbioses. This mechanism is even less plausible than evolution itself. The eukaryotic cell was a critical stepping stone for the arising of higher organisms including man as it enables the construction of complex multicellular bodies.

The functioning of the eukaryotic cell is infinitely more complex than the prokaryotic cell, especially the new structure and properties of DNA. In the new cell we can find several new very important parts such as mitochondria, chloroplasts, nucleus, membrane and cytoskeleton which change the workings of the cell.

The fact that information about the construction and the functioning of the cell is not recorded in the genes begs the question - where is this information stored? It appears that it must be stored in the cell itself. A new cell arises as a result of the division of the old cell. This means that the information in a new cell is an exact copy of the information in the old cell. There is no mechanism to modify this information. If we go back in time to the beginning of life, then there must have existed one cell-mother, the progenitor of all life. How can this be related to Darwin's evolution theory which states that all organisms changed slowly as a result of small genetic mutations?

As discussed in chapter 3, the system of making proteins and molecular complexes also confirms the fact that to make a cell there must already exist a parent cell. It demonstrates that to make proteins, amino acids are needed, which can be made with the assistance of enzymes built from amino acids. The conclusion is that information provided by the genes is not sufficient to initiate life and therefore life must already exist in order to make new life.

So what is the origin of the mother cell? The mother cell could not have arisen as a result of some changes or modifications because her daughters are an exact copy of her. We are talking not only about the complex architecture of the cell but about a very complex control system directing the cell's reproductive and metabolism processes. These processes, because they are not recorded in the genes[4] cannot change, therefore they have been present from the very beginning of the existence of cells. So how could such complex cell structures have existed right from the beginning of life on Earth? From where could thousands of instructions controlling the cell's life and reproduction have originated?

The design of the biological system secures the stability and continuation of life. We already have proof of that because bacteria have been around for at least 3.5 billion years and current complex cells for about 2 billion years. What is interesting is that organisms which existed a few billion years ago are still living on Earth today.

CHAPTER 5

Energy generation and transformation

Energy system

The first life forms which arrived on Earth were single cell organisms. At a perfunctory glance these organisms seem very simple, however this is a very deceptive impression. Inside the simplest cell highly complex molecular machinery is at work. Such machinery is used to generate, store and transform energy.

Life, especially intelligent life, requires a lot of energy, therefore the design of an efficient energy system was critical to sustain living organisms. The most important part of the design was to provide a supply of energy for plants and animals in such a way that the implemented system would be able to meet the future energy demands of highly developed organisms including humans. However, such an energy system had to be introduced right at the beginning of life, therefore even the first simple organisms had to be equipped with very sophisticated energy generation and transformation modules.

This approach to the design can be confirmed by the fact that simple life, for example some bacteria, could rely only on chemical energy for living. However this source of energy, which was abundant and easily available in volcanic waters, while adequate for many single cell organisms would not be sufficient to support plants and animals living on land. Also such chemical resources could eventually be depleted and therefore would not support advanced life over a long period of time. The only viable long term solution was to use the limitless source of energy coming from the Sun. The Sun delivers a huge amount of radiation energy to the Earth's surface amounting[1] to about 1,050 W/m^2 and about 42 percent of this energy is in the visible spectrum.

The Sun's energy has to somehow be extracted simply and efficiently and this is not an easy undertaking. Our present knowledge only allows us to use the Sun's energy for heating water or other liquids or to generate electricity in photovoltaic cells. However none of these technologies are suitable for biological organisms. Therefore a more sophisticated system was needed, initially absorbing the Sun's radiation energy and converting it into chemical energy which could then be stored in an easily reusable form. Once the energy was stored a different biochemical system was needed to extract energy from the chemical storage and change it into appropriate mechanical, chemical or thermal energy. Both systems had to be developed at the same time to be of use.

The process of converting and storing the Sun's energy is called photosynthesis and the process of extracting energy from the stored chemical compounds is called respiration. An overview diagram of the energy system in plant cells is shown in Figure 5-1. An outline of this system will be presented later in this chapter.

Photosynthesis and respiration are equal and opposite processes. Photosynthesis in plants takes place in an organelle* called chloroplasts* and makes sugar molecules from carbon dioxide and water using sunlight as the source of energy. It releases oxygen into the air as waste. Respiration takes place in mitochondria* and does the opposite. It uses oxygen to release energy from sugar molecules and produces water as waste.

The processes of photosynthesis and respiration are very intricate and it would be impossible to explain them fully in a short chapter. Extensive biochemical and bio-molecular knowledge is needed to understand these processes completely. What I will try to achieve in this chapter is to show the extraordinary complexity of these processes which came into existence right at the beginning of life on Earth. To follow these processes there is no need to know biochemistry, but I have decided to use proper scientific vocabulary so the reader would be able to obtain additional information if required.

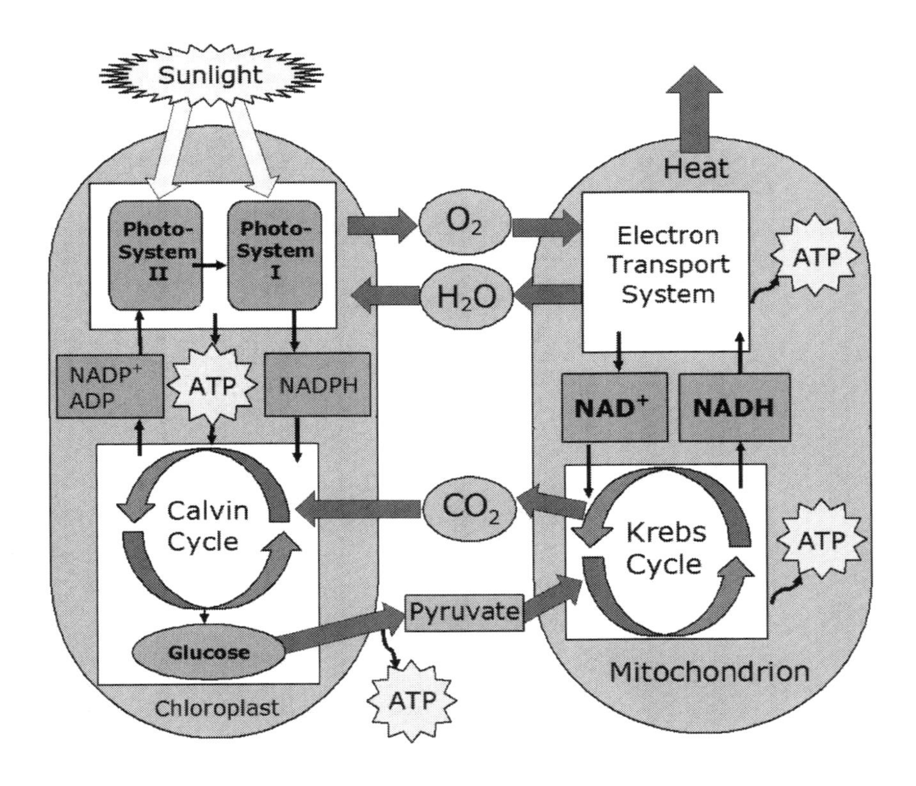

Figure 5-1. Overview diagram for the energy system of plants.

ATP*

The energy system needs a fuel which can be easily produced and used to deliver energy to every cell of every living organism. For life to form a coherent and mutually dependent system there is a need for a single universal energy currency which can perform all energy tasks. The function of this universal currency is fulfilled by an *adenosine triphosphate* molecule abbreviated as ATP. What is interesting is that all organisms from the simplest bacteria, fungi and plants, to all animals, mammals and man use exactly the same type of fuel.

ATP is not a simple molecule and its molecular mass[2] is 507 Da. Splitting off the terminal phosphate group from ATP releases a large amount of energy. This splitting can be expresses by the equation:

$$\text{ATP} \rightarrow \text{ADP} + \text{Pi} + \text{energy}$$

Where ADP* is the *adenosine diphosphate* and Pi is the phosphate group.

Just the cutting of one bond is sufficient to liberate about 30.6 kJ/mol. ADP is recycled again by adding a phosphate group, and then changed into ATP using energy obtained during respiration.

ATP fuels every activity of living organisms and is produced in large quantities. A single human cell uses about 10 million molecules of ATP every second. The production of ATP must meet all energy requirements and could be as high as 65 kg of ATP per day, equal to the average mass of a person, although at any given moment only about 100 g of ATP is in our bodies. In humans each ATP molecule can be recycled up to 750 times per day. All main life processes such as respiration, fermentation and photosynthesis produce ATP.

Photosynthesis and energy storage

Photosynthesis provides energy which powers life on Earth. All the energy released by the burning of coal, firewood, liquid fuels, natural gas and produced by our bodies as a result of burning all the food we eat has been captured from sunlight by photosynthesis. During photosynthesis electromagnetic energy from sunlight is used to make chemical energy using carbon dioxide for the carbon and water for the hydrogen. It is an ingenious process which takes the two most common inorganic materials and makes organic molecules which are used for energy storage.

Photosynthesis general equation may be written as:

$$CO_2 + 2H_2O + \text{sunlight} \rightarrow (CH_2O)_n + O_2 + H_2O + \text{heat}$$

Where $(CH_2O)_n$ is the generic name for energy-rich compounds like sugars and starches. Oxygen, O_2 which is released into the atmosphere comes from the water.

The process of photosynthesis in plants takes place in special organelles called chloroplasts (Figure 5-1). It is convenient to divide the photosynthetic process in plants into three stages, each occurring in a defined area of the chloroplast: (1) capturing energy from sunlight, (2) using this energy to make ATP and NADPH* and (3) using the ATP and NADPH to power the synthesis of carbohydrates from carbon dioxide in the air. The first two stages are called light reactions and the third stage is called the Calvin cycle where light is not required. All three stages of photosynthesis are tightly coupled and controlled so as to produce the amount of carbohydrate required by the plant.

The energy for photosynthesis comes from sunlight. Light is an electromagnetic wave similar to radio waves but having much shorter wavelengths. A wavelength can be visualized as the distance between the two closest peaks of waves, for example sea waves. The Sun radiates energy in a very wide range of wavelengths, from ultraviolet (UV) having wavelengths from 200 nm[3] to 400 nm, visible light (400-700 nm), to infrared radiation (IR) having wavelengths from 700 nm to about 3,000 nm. Light of short wavelengths has high energy and we know that excessive UV radiation can cause skin burns or even cancer, while infrared light having longer wavelengths has low energy and just provides nice warmth. Light travels in short bursts of radiation called photons. Each photon has certain energy proportional to its frequency and can be treated as a particle.

The sunlight is collected by chlorophyll* which is a green pigment located in plant chloroplasts and cyanobacteria*. Chlorophyll molecules absorb light of certain wavelengths. Normally there are several different chlorophylls in plant leaves which absorb blue, violet and red light, and reflect yellow and green light giving them a green color. Chlorophyll is embedded in an extraordinary membrane system inside the chloroplasts. When a photon of the correct wavelength strikes any chlorophyll molecule within a chloroplast, it is absorbed by that molecule forcing one electron out of the molecule. Then a chain reaction takes place during which the excitation energy is transferred from one molecule to another within the cluster of chlorophyll molecules until it encounters the reaction centre and then the electron transfer is initiated. The chlorophyll which lost the electron is highly reactive and takes an electron from the nearest suitable source, which in this case is water. Accepting an electron from water sets the chlorophyll molecule into a stable state waiting for further light to strike.

Chlorophyll has a peculiar structure and essentially consists of two parts: a ring and the long carbon chain as shown in Figure 5-2. The ring with four nitrogen atoms is bound strongly to a magnesium atom in a square planar arrangement. The chemical formula for chlorophyll is $C_{55}H_{72}MgN_4O_5$ and its molecular mass is 893 Da.

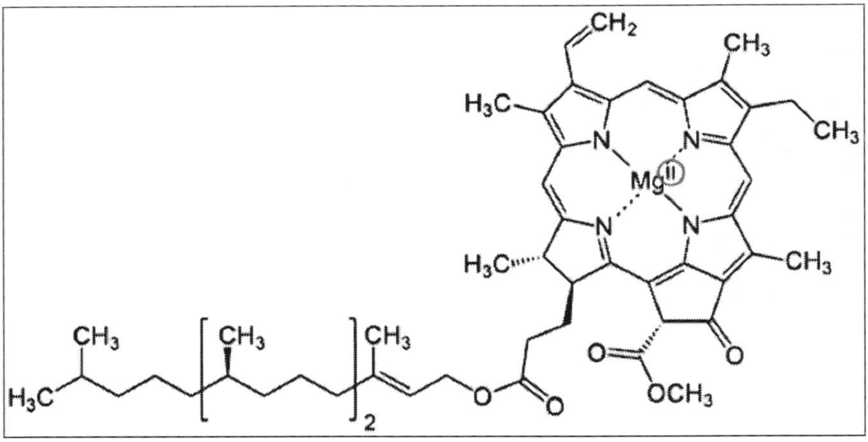

Figure 5-2. Chlorophyll structure

Biosynthesis of chlorophyll is extremely complex. It is difficult to convey the complexity in this book so I have quoted an abstract from the paper *Chlorophyll Biosynthesis in Higher Plants*[4]. You do not have to know all the chemical compounds mentioned to realize how complex the process is.

"Chlorophyll (Chl) is essential for light harvesting and energy transduction in photosynthesis. The Chl biosynthesis pathway in higher plants is complex and is mediated by more than 17 enzymes. The formation of

Chl can be subdivided into four parts: (1) synthesis of 5-aminolevulinic acid (ALA), the precursor of Chl and heme; (2) formation of a pyrrole ring porphobilinogen from the condensation reaction of two molecules of ALA and assembly of four pyrroles leading to the synthesis of the first closed tetrapyrrole having inversion of ring D, i.e., uroporphyrinogen III; (3) synthesis of protoporphyrin IX via several decarboxylation and oxygenation reactions, and (4) insertion of Mg to the protoporphyrin IX (PPIX) moiety steering it to the Mg-branch of tetrapyrrole synthesis leading to the formation of Chl".

Is it possible that such a complex process was "invented" by evolution right at the beginning of life on Earth? The question is where would these intermediate chemical components have come from?

Water splitting process

The second critical function of the photosynthesis process is splitting water into hydrogen and oxygen. The photosynthesis process needs a source of protons[5] and electrons to transform the Sun's energy into chemical energy. This source is water, but water is a very stable molecule and is not easily split. At present we are able to split water using electrolysis and steam reformation, but these processes consume more energy than they could produce by burning hydrogen.

The photosynthesis water-splitting reaction can be expressed as:

$$2H_2O \rightarrow O_2 + 4H^+ + 4e^-$$

where O_2 is the oxygen molecule, H^+ is the proton and e^- is the electron.

The splitting of water is a very difficult process and needs the help of a catalyst which is a part of the so called oxygen-evolving complex. The oxygen-evolving complex contains three proteins: D1 built from about 4,500 atoms, D2 made from 3,300 atoms and D3 consisting of 2,400 atoms. At the heart of the catalyst are four manganese and one calcium atom which are connected to five oxygen atoms and are linked to different amino acids of these three proteins (Figure 5-3).

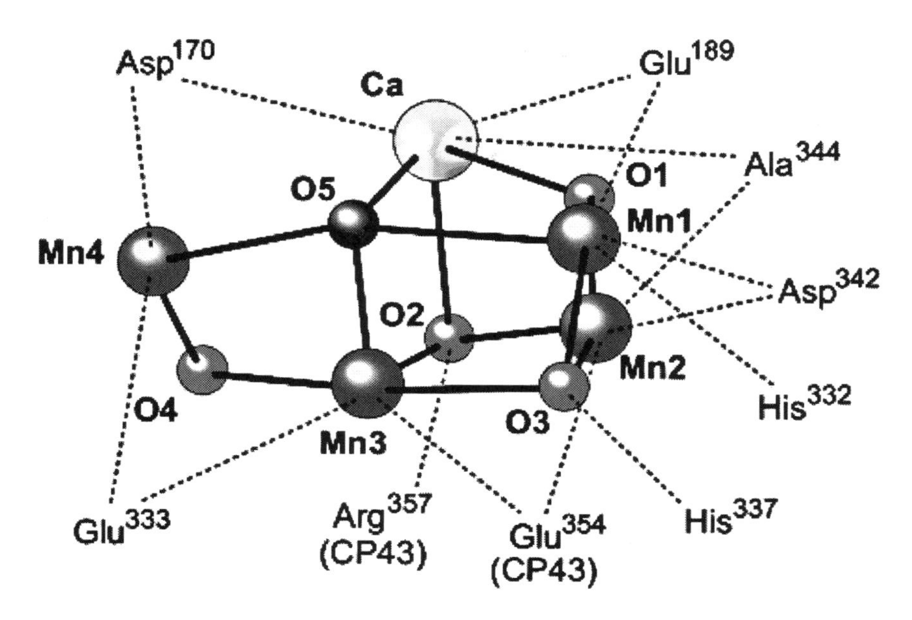

Figure 5-3. Possible arrangement of metal atoms in the water splitting catalyst.

In the oxygen-evolving complex the oxygen atoms of two water molecules bind to a cluster of manganese atoms which are embedded within the complex and bound to the reaction centre. In a way that is not fully understood, this enzyme splits water by removing electrons one at a time to fill the holes left in the reaction centre by the departure of light-energized electrons. As soon as four electrons have been removed from the two water molecules, oxygen is released.

Oxygen exists in our atmosphere due to this molecule. This is a very mysterious molecule which for many years evaded discovery and only recently its structure has been elucidated[6]. The process of splitting water is so complex that it is still not fully understood in spite of the fact that the structure of the oxygen-evolving complex is now well known. Understanding the process has eluded us not from want of skills and effort. If this process were understood it would open up for mankind an unlimited source of energy using just the Sun's energy and water. It would make the oil industry and nuclear power stations redundant. It would offer cheap energy to the poorest nations on Earth. So there is no doubt that we want to discover it, but so far to no avail.

It is interesting to look at how evolutionists explain the arising of this catalyst. Because a similar manganese compound was discovered to be present in the oceans, it was therefore proposed that the oxygen evolving complex 'assimilated' this compound. Yes, true! It decided to pick this compound out of millions of other compounds present in water. How did it know that this is the right compound which would help to split water? Even at present a *Homo sapiens* brain with its billions of neurons is not able to work out how this water splitting works, but evolutionists say that three billion years ago this simple cell knew that it needed 4 manganese atoms with one calcium atom to do the job.

But the cell not only knew that it needed manganese and calcium atoms, it prepared in advance 3 proteins which were going to interact with these atoms. Without these proteins the water splitting process would not work. Therefore its DNA had to develop 3 protein codes not knowing what the proteins were for, because according to the central dogma*, DNA does not receive communication from the outside world, so it was a "trial and error" process. Besides these three proteins there were also the DNA coded transport proteins, sensor proteins and many others which were discussed in chapters 3 and 4. These proteins must have been developed simultaneously, otherwise the water splitting system would not work. How evolution did this has not been explained yet.

Photosystems

The photosynthesis process is carried out by two large molecular complexes known as photosystem I and photosystem II. Each photosystem is a network of chlorophyll molecules, accessory pigments and associated proteins held within a protein matrix on the surface of the photosynthetic membrane. An overview of the photosynthesis system is shown in Figure 5-4. This figure shows a very simplified process but even so it is beyond the scope of this book to describe it. It is not necessary for the reader to fully understand the processes of photosynthesis, only to realize how complex these processes are.

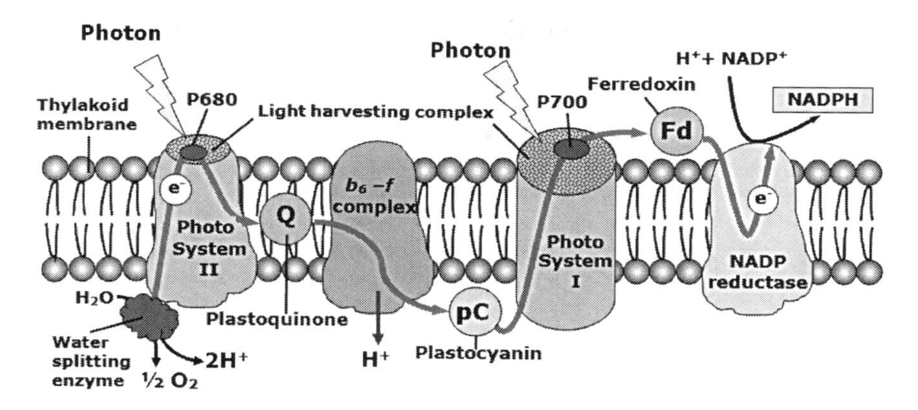

Figure 5-4. Summary of photosynthesis in plants utilizing photosystem I and II.

The main components of photosystem II include the light harvesting complex and the reaction centre with the oxygen-evolving complex. Each light harvesting complex consists of hundreds of chlorophyll molecules which gather light and feed the captured light energy to the reaction centre. In the reaction centre, called $P_{680,}$ there is the oxygen evolving complex and chlorophylls which absorb light having wavelength of 680 nm. In saturating light the reaction centre can have an energy throughput equivalent to 60 MW per mole[7] of Photosystem II which is more than the power of the largest gas turbine.

The complexity of Photosystem II is extraordinary. Only recently, in 2012 using X- ray diffraction, its organizational and structural details were obtained. It includes more than 5,000 amino acids and its molecular mass is about 700 kDa, which corresponds to the mass of about 100,000 atoms[8]. The photosystem II of cyanobacteria and green plants has the dimensions of 10.5 nm depth, 20.5 nm length, and 11 nm width and is composed of around 20 subunits (depending on the organism) as well as other accessories like light-harvesting proteins. Each photosystem II contains at least 99 co-factors*. The description of this complex is beyond the scope of this book.

Next in the photosynthesis chain is photosystem I which was discovered before photosystem II. Photosystem I has a molecular mass of 1068 kDa corresponding to 152,000 atoms. It has about 120 co-factors, three ferrite-sulphur clusters and a 4 sub-unit light-harvesting complex that captures light and channels its energy to the reaction centre. Its light harvesting complex consists of 130 chlorophyll (a) and accessory pigment molecules. Its reaction centre, called $P_{700,}$ which absorbs light having wavelength of 700 nm, consists of at least 13 protein sub-units.

Photosystem II and photosystem I work in sequence. A photon of light ejects a high-energy electron from photosystem II. The energy of that electron is used to pump a proton across the thylakoid* membrane, contributing to the production of a molecule of ATP[9]. Eventually this electron arrives at photosystem I where a second photon pushes it to a higher energy level. In photosystem I this electron is transferred to a molecule of $NADP^+$ changing it to NADPH. For every pair of electrons one molecule of NADPH is generated.

The net effect of the light reactions is to convert radiant energy into free energy in the form of NADPH and ATP. In the light reactions, the transfer of a single electron from water to $NADP^+$ involves about 30 metal ions and 7 aromatic groups.

The third stage of photosynthesis is to make organic molecules from carbon dioxide. This process takes place during the Calvin* cycle where the enzyme Rubisco* binds carbon dioxide and incorporates it into the carbohydrate. Rubisco is very important in terms of biological impact because it catalyzes the most commonly used chemical reaction by which inorganic carbon enters the biosphere. Rubisco is a very large enzyme made of about 70,000 atoms. It is composed of eight large subunits and eight small subunits. During the Calvin cycle carbon dioxide comes from air, the energy needed for the reactions is stored in ATP and NADPH provides a source of hydrogen and energetic electrons. Overall thirteen enzymes are required to catalyze the reactions in the Calvin cycle. The energy conversion efficiency of the Calvin cycle is approximately 90 percent. Although the Calvin cycle does not require light it cannot operate at night because it needs NADPH which is generated by light. Without Rubisco photosynthesis could not be completed. This enzyme works very sluggishly processing only three carbon dioxide molecules per second. Because it works so slowly many Rubisco molecules are needed and in a typical plant leaf 50 percent of all the proteins are Rubisco.

The photosynthesis process is not very efficient. Because photosynthesis only uses red, blue and violet wavelengths of the Sun's radiation, it only actually captures about 1 percent of energy from the solar spectrum. During the photosynthesis process most of the energy captured is lost as heat and only 1 percent or so of the total energy absorbed by leaves is used to produce sugar molecules. Therefore photosynthesis converts only approximately one ten thousandth of the available energy from the Sun. Because of this low efficiency, the food needed to support large herbivore animals must come from vast grassy areas and this is the factor that limits animal populations on Earth. However, over many millions of years the photosynthesis process has generated a huge amount of energy which has been captured in the deposits of coal, oil and gas.

Respiration

Lavoisier, the father of modern chemistry and one of the greatest French scientists, first correctly identified the respiration of animals as a slow combustion process during which carbon and hydrogen are burnt like oil in a lamp. What is the origin of the energy released during respiration? The energy is stored in the bond[10] of molecules such as glucose. When the bond is broken its energy is released. We now know that the combustion of glucose in respiration is an oxidation process during which a substance loses an electron. When sugar molecules are burnt by plants using oxygen, they produce ATP but also carbohydrates, lipids and proteins. During this process carbon dioxide and water are released back into the environment and the recovered energy is used to power life. Cells without chloroplasts, such as animal cells, require an outside source of glucose and oxygen, and like plants, also generate carbon dioxide and water. The oxygen put into the air by photosynthesis is taken out again by respiration. This is how a never ending fine equilibrium is maintained.

Mitochondria

Respiration in complex cells takes place in special cellular components called mitochondria. There can be thousands of them in a single cell where they use oxygen to burn up food. They are so small that a billion of them would fit onto a pinhead. All animals as well as plants and algae contain at least some mitochondria. There are about 10 trillion mitochondria in the human body which constitutes about 10 percent of its mass.

Mitochondria (Figure 5-5) have two membranes. The outer membrane is smooth and continuous, separating the mitochondria from the rest of the cell, and the inner membrane is convoluted into hundreds of folds resulting in a large surface area. Thousands of molecular complexes are embedded in this large inner membrane forming the respiratory system. Space inside the inner membrane is called the matrix which includes DNA, ribosomes and mRNA. The main product of mitochondria is ATP which provides energy to every cell, but the process which generates ATP consists of several stages including the Krebs cycle, electron transport system and ATP synthase. These stages are discussed later in this chapter.

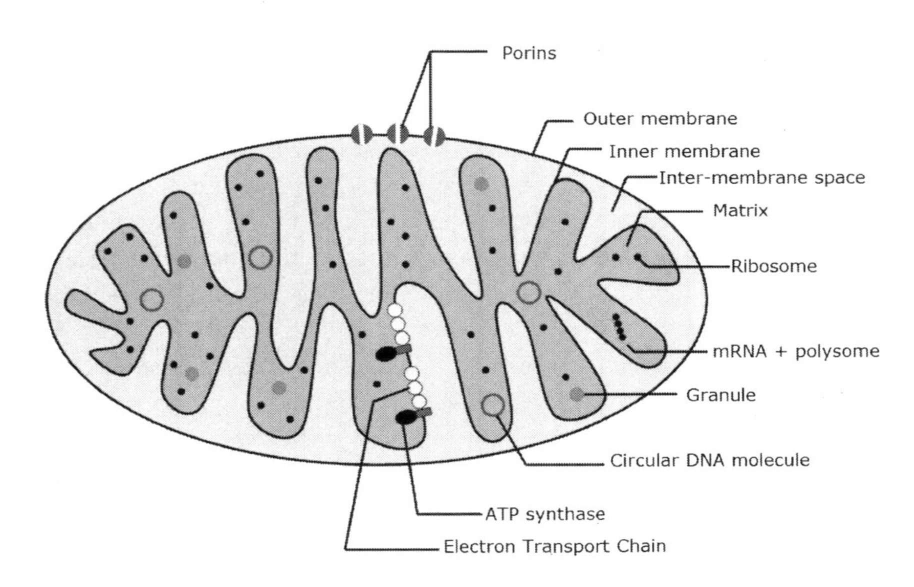

Figure 5-5. Structure of a mitochondrion

Practically all mitochondrial proteins, about 1,500 of them, used in the respiration molecular complexes are encoded by genes in the cell nucleus. There are about 13 mitochondrial proteins, the core units, which are critical for the functioning of mitochondria. These 13 proteins are coded by the genes which are in mitochondria. Because every mitochondrion contains 5 to 10 copies of these genes and because there are hundreds of mitochondria in every cell there are thousands of copies of the same gene in each cell, compared with only two copies of genes in the cell nucleus. So many copies are needed because the mitochondrial genes are so crucial for the continuation of life that they must not be affected by mutations.

The mitochondrial genes pass to the next generation only in the egg cell, not in the sperm. This means that mitochondrial genes pass only down the female line. Because these genes are so critical they do not change or change very slowly, therefore they are used to trace our ancestors on the maternal line back to our common ancestor the 'African Eve', which apparently lived about 200,000 years ago.

*Krebs cycle**

During the first step of the respiratory system, glucose which was obtained from photosynthesis or from other sources is broken down into smaller fragments in the process called the Krebs cycle[11]. The Krebs cycle consumes organic molecules from food and spins off hydrogen needed for generating energy as ATP. This cycle is at the heart of the cell's physiology as well as its biochemistry.

The Krebs cycle is a very complex chemical reaction process and it consists of 9 main reaction stages and 10 minor stages. Carbon and oxygen are stripped down during these reactions and discharged as carbon dioxide waste. The remaining hydrogen atoms are split into their constituent electrons and protons. Then electrons are passed along the electron transport system by the electron carriers. Of all possible organic molecules, those of the Krebs cycle are the most stable and so the most likely to form. During the Krebs cycle a very important compound, NADH, is produced which is used to carry electrons which are fed to complex 1 in the electron transport system.

Electron transport system

While we could call the mitochondrion the engine room of the cell, the electron transport system is the engine which generates energy used by the cell. Without it life would be impossible and if the operation of the system came to a stop the cell would die. The electron transport system consists of four gigantic molecular complexes imbedded in the inner mitochondrion membrane. They use energy released by the flow of electrons along the system to pump protons across this membrane. The respiration system includes the electron transport system and ATP synthase. A schematic diagram of the respiration system is shown in Figure 5-6.

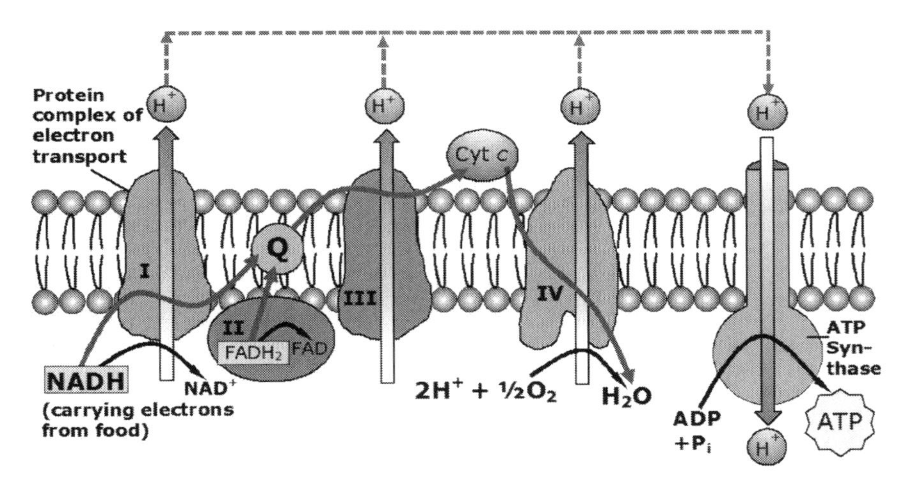

Figure 5-6. Respiration system in mitochondria.

Electrons enter complex I where their energy is used to transfer protons, H^+ through the membrane. Normally a pair of electrons will push 4 protons out on the other side of the membrane. The complex I unit is very intricate (Figure 5-7). It is composed of numerous proteins, coenzymes and cytochrome*. In mammals it comprises of 44 sub-units having a molecular mass of about 1,000 kDa. This means that it is built from about 140,000 atoms. It is not only the size of this complex which is astounding, but its operation is well beyond any man made devices. Whilst we know quite a lot about the workings of this complex the details of the processes are still not fully understood.

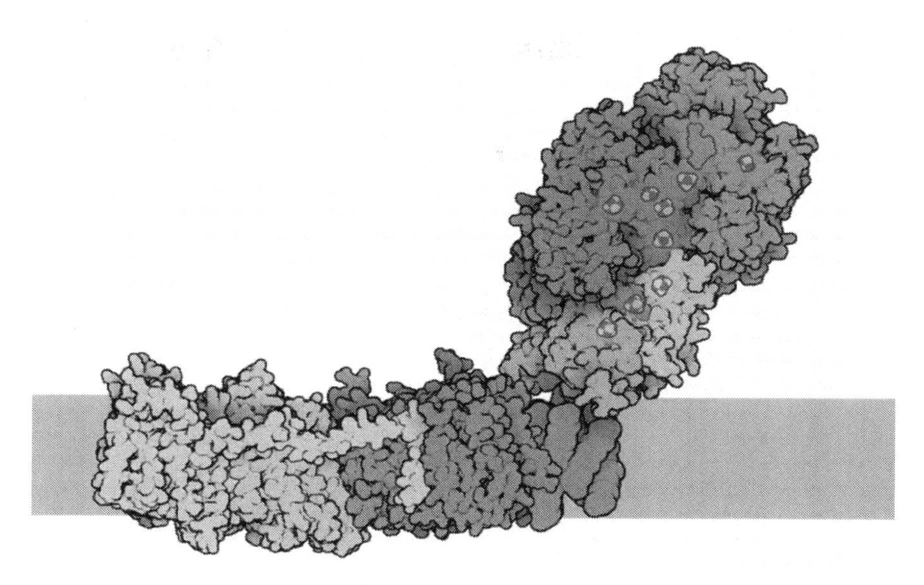

Figure 5-7. Structure of complex I based on X-ray crystallography.

After leaving some energy in complex I a pair of electrons passes to complex III where they thrust another 4 protons across the membrane. Complex II is another entry point for electrons to the chain but it does not transfer protons to the other side of the membrane. After leaving complex III, electrons move to complex IV where they are joined with oxygen to form water. In this last stage they still manage to generate 2 protons. This way one pair of electrons helps to generate ten protons which accumulate on the other side of the inner membrane. These protons can now be used to make the most important molecule of energy - ATP. The mechanism of pumping protons across this membrane is fundamental to the processes of life.

As mentioned above the three respiratory complexes use energy released by electron flow to push protons across the inner membrane. On the outside of the membrane protons accumulate and because they carry an electrical charge an electrical voltage of about 150 mV is built up across the membrane. Since the membrane is about 5 nm thick this results in an electric field gradient of about 30 million volts per meter which is 10 times higher than the dielectric strength* of air. Therefore this gradient could not exist in the presence of any air gaps. As a result of this process a reservoir of protons is built up behind the membrane which simply acts as a dam. These protons are used to generate ATP whenever it is required.

ATP synthase

ATP is produced by ATP synthase having a molecular mass of about 600 kDa. It is built from about 86,000 atoms. ATP synthase uses ADP as a supply material and by adding a phosphate group changes it into ATP.

Up to 30,000 ATP synthase complexes can be embedded in the inner mitochondrial membrane. ATP synthase is an ingenious example of nanotechnology. It is the smallest rotary motor built from protein molecules driven by a proton gradient or electric field gradient, so one could say that it is an electric motor.

Recent research has unveiled the complicated structure of this molecule. Using X ray diffraction techniques the atomic structure of ATP synthase was identified. A schematic representation of this structure is shown in Figure 5-8.

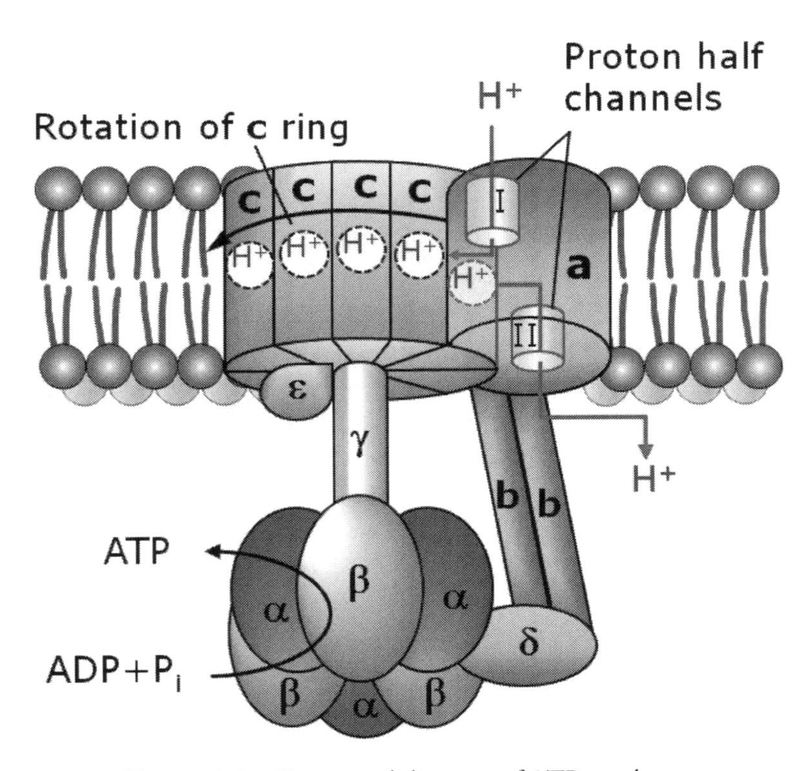

Figure 5-8. Functional diagram of ATP synthase.

The diameter of the ATP synthase in the bacteria E. coli is about 10 nm and its overall height is 20 nm. It consists of several components: the drive shaft (γ) which passes through the membrane; a ring (c) having 10 to 14 segments is attached to this shaft. In each segment there is a cavity for one proton. Protons are fed to the segments via the transfer unit (a) which is fixed to the membrane. The transfer unit has two proton half channels. One half channel (I) has an opening to the outside of the mitochondrion inner membrane where the protons are stored and another half-channel (II) next to it, has an opening to the inside of the mitochondrion i.e. the matrix. The proton enters the half-channel (I), moves through the hole and is transferred to the cavity in the ring (c) when the segment is aligned with the hole. The ring always rotates in a clockwise direction as seen from the outside of the mitochondrion and carries the proton back to the transfer unit (a) where it enters the second half-channel (II) and exits to the inside of the membrane. The ring rotates 25-36 degrees until its next segment cavity is aligned with the half-channel hole (I) and the next proton enters this cavity and so on. It is still not known how protons drive the rotation movement, but there must most likely be an interaction between the proton charges and the proteins.

The shaft (γ), attached to the ring (c), extends outside the membrane inside the mitochondrion. The other end of the shaft is placed in the head (α, β) which is responsible for making ATP. The head is prevented from rotating by the stator rod (b) which has one end fixed to the head via a (δ) unit and the other end is fixed to the transfer unit (a).

The head consists of three alpha (α) units and three beta (β) units in a hexametric arrangement. Only beta units generate ATP. The part of the rotating shaft (γ) placed in the head has a cam profile which switches beta units into 3 states: open, tense and loose. Rotation of the cam by 120 degrees causes the change of state. In the open state the beta unit releases an ATP molecule and accepts a new ADP molecule. In the loose state phosphate is added to the ADP. In the tense state the ADP is changed to ATP, therefore three ATP molecules are formed per revolution.

In humans the ring (c) has 10 segments therefore 10 protons are needed for one turn of the head. The ATP synthase motor, when saturated with protons, can rotate up to 30,000 rpm and is able to produce approximately 390 ATP molecules per second. With its near 100 percent efficiency, far surpassing human technology, ATP synthase shows clear evidence not merely of engineering but of exceptional design abilities. Scientists have been trying to uncover the "secret" behind this very efficient mode of operation for quite some time. Unfortunately, even after more than 30 years of research we still don't fully understand how ATP synthase really works.

Comments

The cell energy system is in reality far more complex than has been described in this chapter. It includes thousands of different proteins, enzymes and auxiliary compounds working together as finely tuned machinery. Many hundreds of chemical reactions take place in this system involving very sophisticated catalysts such as a water splitting enzyme. Some chemical reactions are in the form of loops such as the Krebs and Calvin cycles. These cycles are so precise that it is difficult to even consider that they came about as a result of many random mutations.

The fact that the life energy system is based on electron transport and proton pumping across the membrane is in itself unbelievable. When it was proposed by Mitchell in 1961 nobody believed him and it took almost twenty years to be accepted[12]. This discovery was probably more important than the discovery of the DNA double helix, but because its mechanism was so complex it did not catch the public's imagination.

The pinnacle of bio-technology, without any doubt, must be ATP synthase, such a superb nano-motor made of proteins. Its operation is so intricate that we are still puzzled how it works. Its ingenious design is so accomplished that it could not arise as the result of a blind, random process.

The most important fact to consider is that all this machinery has existed right from the beginning of life on Earth and has not changed since then. What is fascinating is that the first simple cells did not need such an efficient and intricate energy system. This system is only needed for advanced animals and man, but was implemented right at the beginning of life with an understanding of how life would develop. While huge progress has been made during the last fifty years, we still do not know everything about the workings of the energy system and there are many areas which need more investigation.

The energy system in photosynthesis in bacteria, although based on the same chemical reactions and similar molecular complexes as in plants, does not contain chloroplasts. For example in cyanobacteria* there are several internal membranes in which photosystem II and photosystem I are embedded as in chloroplasts and perform the same function.

Bacteria do not have mitochondria, but again have an internal membrane which contain electron transport system complexes. The only difference is that while in mitochondria the internal membrane is folded and convoluted providing a very large surface area, in bacteria this membrane follows the shape of the external cell wall and has a much smaller area, consequently containing fewer electron transport system complexes. As a result bacteria are not able to generate as much energy as mitochondria.

The oxygen respiration process is about 40 percent efficient which is critical in a long food chain where, for example, small insects living on fungus are eaten by larger insects which in turn are eaten by small animals which serve as a food for larger animals and so on. With low respiration efficiency this food chain would come to an abrupt end. For example, for a respiration efficiency of 10 percent, after 3 stages only 0.1 percent of the original energy would be utilized by the final predator meaning that a huge amount of original resources would be needed to support higher life. So oxygen respiration can be the only solution to secure the existence of large carnivorous animals and man.

CHAPTER 6

Cambrian explosion

After the arising of eukaryotic cells about 2.2 billion years ago, life on Earth passed through a long period of stagnation lasting about 1.5 billion years. Although microbial life was well established, no complex organisms were developing and there are few fossils from this period. However, that does not mean that nothing important happened because Earth was going through many changes. The most important was the generation of oxygen by cyanobacteria which was discussed in chapter 3. Oxygen levels were slowly rising, eventually reaching present day levels about 600 million years ago. The Earth also went through several glacial periods from about 750 to 600 million years ago when the whole planet was covered by snow and ice. These periods of instability were the main reason why more advanced life had not developed. Up until then life consisted mainly of single cell organisms living in water and there were no terrestrial organisms. Suddenly, out of nowhere, there arrived complex multicellular organisms. The increase in complexity was just incredible, from a single cell to multi trillion cell organisms. This period is called the Cambrian explosion, because it was like a true life explosion. It started around 541 million years ago in the Cambrian Period and lasted for the next 20–25 million years, during which all major animal phyla appeared as indicated by fossil records.

Phylum

To appreciate the importance of this event we have to know some basic facts about the classification of life. All life on Earth is traditionally divided into 5 kingdoms: animals, plants, fungi, bacteria and protista[1]. Each kingdom is divided into groups called phylum, and animals are classified into 32 phyla. Of these, 20 are minor phyla, and at least 21 are exclusively aquatic, with several others in quasi-aquatic habitats on land. None are entirely terrestrial showing that water was the main habitat for early life.

Informally, phyla can be thought of as a group of organisms based on similar body plans. The most common phyla are:

Phylum Chordata – all the animals which have a backbone such as fish, amphibians, reptiles, birds, and mammals including man.
Phylum Arthropoda – all the 'jointed legged' animals such as insects, arachnids, and crustaceans. All of these animals have an exoskeleton, meaning the skeleton is on the outside of the body. Trilobites belonged to this phylum.
Phylum Mollusca – soft-bodied animals that sometimes have a hard shell. Includes: snails, slugs, octopus, squid, clams, oysters, and mussels.
Phylum Cnidaria – soft-bodied, jelly-like animals with tentacles and venom glands. Includes: hydra, jellyfish, anemones and coral.

We can see that phylum is a very broad category and encompasses very different classes of animals. However, the major body plan is the same and for example almost all animals belonging to the phylum Chordata have not only a backbone but also a distinguished head with jaws and eyes, central nervous system, digestive system and gills or lungs. Altogether there are about 64,000 species belonging to this phylum. So, although men look so different, for example, from reptiles, their basic structures and internal organs are very similar.

Precambrian period

Dating of the Cambrian explosion is based on the discoveries of fossils, indicating that during that time there must have already existed many specimens and colonies of various phyla. However, the principles of multicellular designs had to be developed much earlier. The first multicellular organisms appeared 25 million years before the Cambrian explosion in the Ediacaran period. Ediacaran fauna is a unique assemblage of soft-bodied organisms preserved as fossil impressions in sandstone at the Precambrian time. The Ediacaran fauna, named after the Ediacara Hills of South Australia in which they were discovered in 1946, were the first animals made up of more than one type of cell that required atmospheric oxygen for their growth.

The first fossils of large bodied animals called vendobionts were similar to that of jellyfish measuring up to a meter or so across. They did not have a digestive tract and their body structure looked like a pack of tubes linked together. Vendobionts were very primitive and they could not be classified as the ancestors of later Cambrian animals but they were an early experiment in multicellular life. Beings creating new organisms did not have the experience of how to design them and as we know in engineering there is "a learning period" during which experiments are carried out and prototypes are built. During this period the prototype design was not to be permanently implemented, but provided the experience needed for further improved designs.

The vendobionts were not the only inhabitants during the Ediacaran period. Tracks of small worms were preserved in the sandstone of Namibia, Australia and Siberia. These tracks were left by worms, a few centimeters long, which burrowed in mud on the sea floor. The bodies of these worms were much more complex than that of the vendobionts and had muscles, circulation of fluids carrying oxygen, some kind of a pumping system, digestive system and to coordinate their movement they must have had a simple nervous system. So although we think of worms as being very primitive animals, they have many organs that are also used by more advanced animals. It seems that the Ediacaran period was a testing ground for new body plans. Animals from this period were not that well designed to adapt to varying conditions and they soon became extinct. But as a result of this exercise new experience had been gained and it paved the way for the next phase – preparation of all present-day major body plans.

Cambrian development

To understand the expanse and importance of the Cambrian development we have to envisage how the Earth looked before this period. We believe that it was covered by vast oceans with land mass concentrated in the southern hemisphere[2]. The total area of the land was similar to the present-day area but it was lumped as one continent. The land was barren and lifeless although there were lakes, rivers and streams. The landscape was dominated by the reddish and brown colors of metal oxides. All life was concentrated in sea water where most of the organisms were single cells such as algae and bacteria forming slimy colonies. It was not a very attractive picture. The oxygen level was high, approaching present day concentrations but it was not used for respiration by higher organisms.

Suddenly the whole picture changed when animals with completely new body plans came into being. We are not talking about two or three different animals but the arising of dozens of phyla. We know about this from fossil records, the most abundant of which are in the Burgess Shale Formation in the Canadian Rockies of British Columbia. It is famous for the exceptional preservation of the soft parts of its fossils. At 508 million years old, it is one of the earliest fossil beds containing soft-part imprints. The fossils found include a wide range of organisms. Free-swimming organisms are relatively rare, with the majority of organisms being bottom dwelling - either moving about or permanently attached to the sea floor. About two-thirds of the Burgess Shale organisms lived by feeding on the organic content in the muddy sea floor, while almost a third filtered out fine particles from water. Under 10 percent of organisms were predators or scavengers therefore having a larger mass.

The body plans of some animals that arose at this time show an ancestry to modern animals. There were also some weird and wonderful creatures unlike any alive today. The ocean teemed with worms, jellyfish, trilobites and brachiopods. Many animals had tough body parts such as shells or exoskeletons which were more easily preserved as fossils. What is most important about this period is that body plans which exist today originated then. It is difficult to argue which group of animals appeared first when all of them came into existence during the same period.

Since the Cambrian explosion millions of new species, but not a single new phylum, have come into existence. It seems that a major design era was completed and since then only lesser changes have been implemented. This is confirmed by the fact that there are no phyla that are exclusively terrestrial, which means that no significant development took place after this period.

What is even more interesting is that during the Cambrian explosion there were more phyla than there are at present, some scientists estimate as many as a hundred. This means that many phyla had an inferior design and became extinct but have not been replaced by any new designs.

Trilobites

Whilst many new animals arose during the Cambrian explosion none of them caught our imagination more than the trilobite. Trilobites were marine invertebrates that had hard, three-lobed shells (the axial lobe and two side lobes), hence the name "trilobite". Their body had three distinctive sections: the head, the chest made of a group of up to 30 segments, and the tail. Underneath were three pairs of legs for the head and paired legs for each segment. They ranged in length from 1.5 millimeters to over 72 centimeters. Paleontologists have identified more than 20,000 different trilobite species, an amazingly diverse group of animals. They all lived in the sea, some burrowed in the mud, some crawled on the surface of the seabed and others swam about in open water or inhabited reefs.

They were protected by a calcite exoskeleton plate which covered the upper surface of the trilobite and curled around the lower edge. The head was covered by a shield and had a pair of eyes and antennae. Although some types of trilobites were blind, most had well developed eyes with very sophisticated compound lenses that had a great depth of vision. Some species had eyes on stalks and these are believed to have buried themselves in the mud, with only their eyes sticking out like periscopes.

The first appearance of trilobites in fossil records was in the Early Cambrian period about 521 million years ago and they finally disappeared in the mass extinction at the end of the Permian period about 250 million years ago. Trilobites were among the most successful of all early animals roaming the oceans for over 270 million years.

Increase in complexity

Organisms living before the Cambrian explosion were mainly single cell bacteria and eukaryotic cells belonging to the kingdom Protista. They were relatively simple, having from 1,500 to several thousand different proteins. The largest bacteria were nitrogen fixing bacteria with more than 7,500 various kinds of proteins and enzymes.

In the Cambrian explosion there arrived, for the first time, multicellular organisms not in the form of a colony of the same cells but organisms with differentiated functions. Hence we witness the appearance of different types of cells, each performing separate specialized functions. These specialized cells had common structures such as a nucleus, mitochondria, and a common energy generating respiration system. But besides this, they had new structures which required new materials. Even the simplest animals needed many new organs such as skin, a skeleton, muscles, neurons, a digestive system, reproductive system, respiratory system, etc. Besides these organs, more advanced animals had eyes, teeth, a brain, a spine, sensors and a complex digestive system.

To provide these functions many new cell designs were needed. Different types of cells had specific jobs to do. Each cell had a special structure that was suited to its job. Several cells that do the same job form body tissue, such as muscle, skin or bone tissue. Groups of different types of cells make up organs such as the heart, liver or lungs. A group of different organs working together make up a system such as a circulatory system or digestive system. So with new body plans we had many new systems needing many new types of cells. For example, mammals use about 250 different types of cells in their bodies. Each type of cell might need hundreds of different proteins, so the new organisms would have needed tens of thousands of new proteins. We know that mammals can have up to 100,000 various types of proteins.

So during the Cambrian explosion we not only saw different body plans, but on a molecular level there was an upsurge in the number and complexity of new proteins, enzymes and cofactors. As a result of this the size of the genome was considerably increased. These new components had to arrive on the scene practically at the same time because they all had to take part in building a body in accordance with the new plans. Even the earliest fossils do not show 'half cooked' animals, or intermediate steps as they are called by evolutionists, but creatures already fully functioning and adapted to the new environment.

Digestive system

On a system level we have the development of a new digestive system, especially evident in scavengers and predators. Up to now the common food was a product of photosynthesis and consisted of mainly carbohydrates, especially sugars. With the development of complex bodies their cells now included many new proteins and fats. Predators must be able to utilize most of their prey's bodies as nourishment, but some of these proteins and fats could be very difficult to digest. Therefore, a new multi stage digestive system had to be developed with a new range of digestive enzymes. By looking in to the present day digestive system of fish we can get some clues as to how complex this new system was. Surprisingly, the basic components and operation of the fish's digestive system is very similar to ours. The digestive system, in a functional sense, starts at the mouth, with the teeth used to capture prey or collect plant foods. After being swallowed food passes to the stomach and intestines where it is digested. But fish already have the two most important digestive organs: the liver and pancreas, providing a variety of digestive enzymes. Whilst the first vertebrates might still not have had a system as complex as ours they already had all the major organs. It was a completely new arrangement never seen before the Cambrian explosion, and what is more important is that this arrangement is still functioning today without any major improvements.

Circulatory system

The increase in size of bodies made it difficult for cells to receive nutrients and oxygen and to remove waste products and carbon dioxide. Multiple layers of millions of cells needed a more efficient transport system and therefore there was a need for the circulatory system.

The circulatory system consists of the heart, arteries, capillaries and veins. It is in the capillaries that the interchange of oxygen, carbon dioxide, nutrients, and other substances such as hormones and waste products takes place. The capillaries lead to the veins, which return the venous blood with its waste products to the heart, kidneys, and gills or lungs. The heart undergoes rhythmic contractions and receives venous blood. The blood passes to the arteries of the gill/lungs and then to the gill capillaries. There, waste gases are given off to the environment, and oxygen is absorbed. The oxygenated blood enters the arteries of the gill arches and from there blood is distributed to the tissues and organs of the body. The circulatory system is not limited to vertebrates but was needed in simple form in all multicellular organisms including earthworms and insects.

Control system

Another important development which did not exist in single cells was the arising of the brain and nervous system. Single cell organisms, even those living in colonies, did not need any information processing facilities. Communication between the cells in the colony was very primitive and used chemical signaling, but these cells did not have to coordinate their functions. In multicellular organisms, cell functioning must be properly synchronized. Activities such as vision, movement, food gathering, reproduction and digestion involve millions of cells and therefore their operation must be coordinated. Such coordination required a very sophisticated control system which had to be developed. This system linked sensors such as vision, touch or smell by sending their signals to the brain which processed them and generated operating instructions. So now we are talking about signal conducting wires, signal processors and activators. Therefore, new components were needed to transmit signals in the form of electrical impulses and new structures were needed to process them. This type of system had never been tried before and had to be built from scratch. Most of the animals that originated in this period had some kind of a nervous system and some of them, such as trilobites, could have had a very sophisticated brain. We know that some of these animals were predators and they needed a much better brain than passive harvesters of food or even scavengers. So this was the start of an information processing revolution which ended with the human brain.

Gene control

Without any doubt one of the most important features which enabled the development of higher organisms was gene switching. The genes for many proteins in cells are not active all the time but must be turned on and off at some point. In a gene there is a control region to which the RNA polymerase must be attached to start gene transcription. When the control region is occupied by a molecule called a repressor, the polymerase is physically prevented from working. To turn on the gene a special molecule, called an activator, binds to the repressor and removes it from the gene which allows polymerase to latch to the control region to start transcription. How the activator "knows" which gene to turn on is not certain.

In higher organisms there are millions of cells which must be arranged in a specific order and in specific positions. To build an organ, many different kinds of cells must be positioned very accurately with relation to other cells. The control of location of these organs is provided by a special set of genes called Hox genes. These genes are essential during embryo development where they control the building of specific parts of the body. The products of Hox genes are Hox proteins. Hox proteins are transcription factors which are capable of binding to specific nucleotide sequences in the DNA called enhancers where they either activate or repress genes. The same Hox protein can act as a repressor in one gene and an activator in another.

Hox genes do not provide instructions on how to build body fragments but control other genes responsible for the assembly of specific body parts. They give information about the location of clusters of cells which form a specific organ. They operate on several, pyramid like levels. On the highest level they provide overall instructions describing the assembly of the whole body. On lower levels they provide instructions for organs and subassemblies. On the top level there is one gene which switches other genes on the level below, and these genes in turn switch further genes on the next level and so on.

Operation of the Hox genes is very complicated and it would be easier to understand their functions by comparing them to car assembly instructions. A series of Hox genes provide instructions on where to put certain parts for the car assembly. The first gene describes the position of the bumper, the next gene describes the position of the radiator, the next gene the engine, then the gearbox, steering wheel, seats, luggage compartment and the rear bumper. There are also gene side branches which describe the positions of the wheels, side doors and windows.

The Hox genes are typically found in an organized cluster in the chromosome. The linear order of the genes within the cluster is directly related to the order of the body regions they affect. For example, in the *Drosophila's* chromosome genes are arranged in the same order as the segments of the fly. So the genes controlling the head are on the left followed by the genes for the thorax and ending with the genes for the abdomen on the right.

If in our car assembly example we replace the bumper instructions with the steering wheel instruction, as a result of this switching the car would have a steering wheel in the front bumper position. And the same happens in nature. When, for example the *Drosophila's* antennae genes were replaced with the limb genes, the fly had legs sticking out of its head. It has been shown in many experiments that mutations in the Hox genes could cause body parts and limbs to be located in the wrong place along the body.

The Hox genes are so important that they must not be changed[3]. The functional conservation of Hox proteins can be demonstrated by the fact that a fly can function perfectly well with chicken Hox proteins in place of its own.

What is even more astounding is that the Hox genes in humans have similar counterparts in fruit flies such as the above mentioned *Drosophila*. But the similarities do not end there. For example, the Hox gene that controls the development of the human head is similar to the Hox gene that controls growth in the head of the fruit fly. And the genes controlling the lower parts of humans are comparable to the genes controlling the tail end of the fly.

The incredible increase in gene control complexity is mostly visible during human embryo development, where about 3,000 gene control substances are used. While Hox genes provide the overall control of embryo development, in practice every cell has to know its location and to be of the right type. There are two general models for how cells are differentiated and form required structures: use of a morphogen gradient, and sequential induction.

Morphogen gradients are generated by a reference cell, normally at the edge of the cell cluster. Such cells produce morphogens - chemical substances which form a concentration gradient[4] along the line of the cells. Cells in the cluster detect the level of morphogen concentration which triggers cell differentiation and this way establishes the positions of the various specialized cell types within a tissue.

During sequential induction a cell generates a substance which affects the differentiation of neighboring cells. For example, the development of the retina induces the development of the lens and cornea of the eye. The substance secreted by the developing retina can only diffuse a short distance and affect neighboring cells, which become other parts of the eye.

The interaction between different systems must be secured by special sets of control genes. For example, we know that human vision requires very extensive signal processing which takes place in the retina and in the optical cortex. Therefore, the eye must interact with the brain and this is reflected at the embryonic development stage. The genes which control the development of the eye are also responsible for the development of the front part of the brain.

Comments

The Cambrian explosion was the period in history which initiated the development of intelligent life on Earth. During these twenty million years, extraordinary development processes took place which included not only the introduction of new body plans for all subsequent animals, but also the development of new multicellular organs with completely new functions and the development of extensive new molecular structures. Many new types of cells came into existence. Cells that were needed to build different organs like muscle, skin, bone and liver, etc. But the most important development was the body plans of all animals, and with this the development of instructions on how to build these bodies by controlling genes at different stages of the development process. The Cambrian explosion stage completed all the grand scale design paradigms, and all subsequent periods focused on smaller design changes at the lower groups of animals such as classes, orders and families.

CHAPTER 7

The conquest of land

All organisms in the Cambrian period were sea based and needed water for their survival. Normally 50 to 90 percent of body mass is water, so living in the sea was the optimum solution for life. Water was a very benign environment where living organisms were protected from ultraviolet radiation.

However, increased oxygen levels helped to generate an ozone layer around the planet which began to provide protection against ultraviolet rays. So by the end of the Cambrian period the Earth was ready to support life on land. However, moving to land was not a natural choice for organisms because it involved significant changes to their body structure and functioning without any striking benefits. Despite this, first the plants arrived followed shortly by animals. So why should life have moved on to dry land?

Plants

The precursors of water borne plants were green algae which harvested the sun's energy using photosynthesis. They thrived in water because, as we discussed in chapter 2, water is transparent to the sun's visible rays needed for photosynthesis. Water provides the perfect environment for plants because it supports their body through buoyancy, provides an unlimited supply of water, enables plants to move with sea currents, facilitates their sexual reproduction and provides mineral nutrition dissolved in water. So the structure of water based plants is relatively simple.

The arrival of plants on land is in itself an incredible story. Whilst we can envisage that animals with legs living in shallow waters could crawl out onto a beach, there is no way algae or other water plants could have done the same. But about 470 million years ago they somehow did, probably with the help of sea tides.

Land was an alien and hostile environment, and in order for plants to survive they had to adapt to the new landscape. They needed completely different body plans to their water based cousins. The light, water, and carbon dioxide they needed for photosynthesis might have been easily accessible on land, but other nutrients for making proteins would not have been available because the soil was sterile. It is quite likely that plants would not have been able to move onto land without symbioses established with fungi and bacteria. This would have helped them get the materials they needed, such as nitrogen, which they ordinarily would have found in the water around them.

The first problem of living on land was the intermittent water supply. Water on land was delivered by rain which could be unpredictable and infrequent, therefore plants would need water retention facilities and protection against dehydration. All plants that live on land have adapted to dry conditions through the development of a waxy cuticle[1] to prevent drying out, structures to absorb and transport water throughout their bodies, and rigid internal support to remain erect without the buoyancy available in water.

The first plants on dry land did not have roots and absorbed water through their outer layers and transferred it to adjacent cells by direct contact. These plants were similar to present day mosses and had to live in wet areas. They lacked strong structural tissues, therefore to support themselves in air they used an ingenious system based on the water pressure inside their cells. The principle of the operation of the plant's supporting structure can be compared to an inflated bouncy castle, which when pressurized has a very rigid structure. In the case of the plant, air is replaced by water which 'inflates' the cells. Since the cells are linked together this results in a firm and rigid assembly.

However, large land plants needed a different system to transport fluids within the plant and to provide support for their tall structures. For the transport of fluids they used special micro pipes creating a vascular system.

This system consisted of two types of vascular tissues called xylem* and phloem*. Xylem transports water and solutes from the roots to the leaves, phloem transports sugars and nutrients from the leaves to the rest of the plant. Xylem has a tubular shape with no cross walls which allows a continuous column of water to flow upwards. The flow takes place through capillary action discussed in chapter 2. When the water in leaves evaporates, more water is pulled from the roots.

Phloem has an elongated, tubular shape with thin-walled tubes. It is composed of living cells that transport sap. The sap is a water based solution rich in sugars and other ingredients made by the photosynthetic areas. Phloem is placed on the outer part of the tree where sap can be collected. Well known applications of sap are the making of sweet syrup from maple trees and latex from rubber trees.

The vascular system was a giant step in the body development of plants. It provided a dual flow system similar to the way blood flows in animals. The main difference is that in plants sap does not need red blood cells because it does not carry oxygen to the cells. All energy processes and transformations take place in the leaves which contain chlorophyll enabling the production of sugar.

However to generate energy, non photosynthetic cells need oxygen which is not transported by phloem. So these cells have to absorb oxygen directly from the air through special openings and as a result of the respiration process they have to exhaust carbon dioxide back into the air. This arrangement limits the growth of plants to the outside cells.

Rigidity of plants is provided to a large extent by the cell wall, which is composed of cellulose, a complex carbohydrate, and lignin, a phenolic compound that stiffens the cellulose fibers. Put together, these features allowed plants to grow much larger and considerably reduced their dependence on moist habitats.

The plants reproductive system had to be adapted to function on land. Plants during their reproductive cycle produce eggs and sperm. When water is present, sperm swim and fertilize the eggs. Therefore plants living on dry land must have such a construction so as to guarantee the presence of water needed for fertilization.

These new features had to be operational from the beginning, so we are not talking about slow adaptation. First, plants had to be protected against drying out (desiccation) in the air, able to absorb nutrients from the soil, grow upright without the support of water, and reproduce on land. Each of these systems required substantial changes in the body plans and functions of plants. When we consider all these impediments it is amazing how big the leap was for plants in moving from water to land.

Animals

When animals moved from water to land they would have had to face similar problems to plants. To cope with these problems, land animals needed a new type of protection such as an outer layer. In early land animals, which were arthropods such as millipedes, the control of water loss was achieved by the exoskeleton having a composite multilayer structure in which the outermost layer was waxy and waterproof. This waterproofing was achieved by a complex system of internal canals providing wax to the outer surface.

Another source of water loss in organisms is excretion. When animals digest proteins, excess nitrogen is produced and is usually released in the form of ammonia which is a toxic compound that requires swift dilution or removal. Dilution demands a lot of water and is thus unsuitable for terrestrial organisms. Thus arthropods converted ammonia into a more complex yet less toxic compound such as uric acid which was often precipitated in solid form and removed with the aid of special organs.

Another major issue with life on land is that of respiration, or gas exchange. This cannot be done through increasing the permeability of the exoskeleton, which would cause significant water loss. The solution to this problem is that arthropods had specialized gas exchange structures which had a very complex composition.

Another less apparent issue is that of reproduction. In the marine environment releasing eggs and sperm into the sea for external fertilization is a workable solution. This is not possible on dry land, therefore internal fertilization was necessary. This required a wide range of complex sperm transfer techniques for safe procreation.

The next problem to overcome was gravity. In water, bodies are supported from all sides by water pressure therefore they do not need strong skeletons to hold internal organs. To swim in water, animals need some sort of fins which do not support the animal's weight, so the muscles used for propulsion do not have to be very strong. Another advantage of living in water is buoyancy which reduces the weight of the body, therefore animals use limbs more for movement than support. To control buoyancy, fish have an air bladder which can be expanded or contracted. This helps fish to stay at a certain depth without using muscle power.

Bodies of land animals are normally, with a few exceptions such as worms, snakes and snails, supported by legs which must be strong enough to carry all their weight. Such legs must be jointed with a strong skeleton and must have powerful muscles. To move on land, animals use much more energy than to swim in water therefore their metabolism must be improved. One can see how effortlessly heavy sea animals such as seals and walruses, which have flippers instead of legs, swim. However, on land they move with difficulty. The largest animal on Earth is a whale because only an aquatic environment can support such a large body. But we realize how weak they are when we watch distressing pictures of whales accidentally beached.

Another problem connected with living on land is ultraviolet radiation. We know that strong radiation can damage DNA and cause mutations which could result in cancer cells. In spite of the ozone layer, which was probably established before the Cambrian explosion, animal's bodies, especially the eyes, must be protected against such radiation. For example, early man had dark eyes where dark pigment protects against excessive light. Blue eyes are the result of a harmful mutation which was less detrimental for people living in northern areas.

The bodies of most animals function in a limited temperature range, say from zero up to 40°C. Land masses are much more exposed to large temperature variations than water therefore land animals must have developed additional temperature control systems. Cold blooded animals would have had a problem living on land, exposed to the cold at night and the sun during the day. They would have needed a hiding place which could provide some protection against the cold and heat. However, only warm blooded animals which arrived much later coped with such a wide temperature range. They had a sophisticated temperature control system which provided heating and cooling of their bodies enabling them to live on land.

The first animals appeared on land about 430 million years ago, but what is interesting is that their body plans which enabled them to do so were generated during the Cambrian explosion about 100 million years earlier.

Why move to land?

When we look at the above list of problems we realize that the transfer to land was not as simple an event as some evolutionists would have us believe. It was not the case that some animals with legs just walked out of the water because they fancied a change. Such a transfer required the coordinated changes of several body structures, organs and behavioral patterns.

So what was the driving force which moved animals onto land? It is difficult to find any biological arguments supporting such a drastic change of environment. Water was a natural supportive environment full of bacteria, protista and small organisms which served as rich nutrients. It was a protective environment characterized by small ambient temperature variations and space-wise it was several orders of magnitude larger than the surface of land. It offered varying habitats from shallow to deep water. In spite of the many disadvantages, life was pushed onto the arid land without any apparent reason.

However, there is a very strong reason. Life in water could only develop up to a certain level. One of the most intelligent mammals, the dolphin, which represent the highest etalon of sea life, lives in water. And yet, for their brains and bodies they need a lot of oxygen and therefore they cannot stay underwater for more than 15-17 minutes. They also have difficulty sleeping, shutting off only half of their brain.

So more developed beings such as humans would have to live on land not only to have a sufficient supply of oxygen, but also to develop tools and eventually civilization. It is difficult to imagine how civilization could have been created by water bound creatures. For example, water is not the right environment to develop metallurgy or electronics. High temperature sources are required to make metals, which of course are impossible to place in water. Sea water being a good conductor is not good for electronic devices. We could say that water is not a practical environment to develop industry, art or science. So it was necessary to push life onto land as a long term plan. This push not only involved animals but also plants which played a crucial role in changing the Earth's surface.

CHAPTER 8

Man

Without any doubt man, and his brain, are the pinnacle and ultimate objective of creation. Earth was preparing for the arrival of intelligent life for 4 billion years. The last 500 million years of the development of life was focused on this objective. The first important step in this direction was the appearance of life on land which changed the Earth's environment and eventually made it habitable for human beings. As has already been mentioned, man's brain needs large quantities of oxygen taken directly from the atmosphere. Whilst water is the perfect environment for life, it is the land which provides the right conditions to develop civilizations with their culture, arts, sciences and technology.

Preparation of the land was a long process and plants played the leading role in this task. The first plants changed the soil and made it suitable for flowering plants. Plants provide an essential contribution to support life. They are the main source of nutrition for man and animals. Animals, in turn, also became a source of food for human beings. Plants provided a source of energy used for heating and cooking, and timber was, and still is used as the main building material for shelters and houses. Therefore, plants paved the way for the arrival of man.

The arrival of modern man is still shrouded in mystery. It is now accepted that *Homo sapiens* appeared about 200,000 years ago in Africa out of nowhere and spread from there across the world. From that time the progress of man's development is more or less well documented. However, the development of early man is much more convoluted and is subject to many controversies.

Early man

How much do we know about the development of early man? Below I try to provide a concise summary of current academic knowledge based on archaeological findings and fossil records. However, the interpretation of these records is challenged by some researchers.

Fossil records of the first hominins*, upright walking apelike beings, are dated to about 4 million years ago. The earliest known hominin, belonging to the *Australopithecus* genus, stood 120 cm to 140 cm (3'11" – 4'7") tall and had a cranial capacity[1] of 300 to 600 cubic centimeters. Several different groups of australopiths lived in East and North Africa between 4.2 and 2 million years ago. Hominins classified as *Australopithecus* did not belong to a homogenous group but were divided into several species such as *A. anamensis* in Kenya, *A. afarensis* in Tanzania, *A. africanus* in South Africa and *A. garhi* in Ethiopia. These species differed greatly anatomically and were on different levels of development. However, there is no proof that there was a link between these groups. Information about these early men is very meager and frequently their fossils consisted of only a small part of the skeleton. These species persisted until not much more than 2 million years ago and then disappeared without a trace. It is most likely that there were more species of australopiths that existed during this period. Furthermore, we don't know how long each of these groups survived. Nevertheless, even if on average, species longevity was only a few hundred thousand years, it is clear that from the very beginning the continent of Africa was host to multiple kinds of hominins.

About 2.5 million years ago, the next step in man's development was the appearance of the genus *Homo* which was more closely related to *Homo sapiens* than *Australopithecus*. The earliest fossils from this genus were found in eastern, south-eastern, and southern Africa. Three species living in these areas were *H. rudolfensis*, *H. habilis* and *H. erectus*.

The earliest known species of early *Homo, H. rudolfensis lived* in Kenya, Ethiopia and northern Malawi between 2.5 and 1.8 million years ago. *H. rudolfensis* was characterized by a large cranial capacity, around 750 cubic centimeters, large molar teeth and a long face that was broad across the eye sockets and flattened below the nose. In several aspects it was similar to the australopiths.

Homo habilis, which dates to between 2.1 and 1.5 million years ago is named from the Latin term ('habilis') meaning "handy" because he was able to use tools. It is one of the most important species in the genus *Homo*. Fossils of *H. habilis* have been found in Tanzania, Ethiopia, Kenya, and South Africa. While this species is distinct from the australopiths in many aspects of its skull, it also exhibits many primitive traits shared with species in the genus *Australopithecus* which suggest it was more similar to its australopiths ancestors. Compared to australopiths, *H. habilis* had a bigger brain, around 680 cubic centimeters, and a more vertical forehead. Additionally, the face and jaws of *H. habilis* were smaller and less projecting than those of the australopiths.

The oldest *Homo erectus* fossils are dated to roughly 1.8 million years ago, while the youngest fossils assigned to this species date to about 300,000 years ago. Remains of *H. erectus* are found throughout Africa and in western and eastern Asia as far as Java. African *H. erectus* is very often called *Homo ergaster.* It was the first fossil, assigned by some scholars to this species, to have been found in Europe, as far north as England. The average brain size of *H. erectus* is estimated to have been approximately 900 cubic centimeters. The braincase and the face and jaws of *H. erectus* were very heavily built, with thick bones. For instance, the brow ridges were massively built and continuous across the face, and large bony prominences existed in the back of the skull. *H. erectus* used more sophisticated tools than other Homo species.

The next step in man's development was the arrival of *Homo heidelbergensis* that lived in Africa, Europe and western Asia between 600,000 and 200,000 years ago. The skulls of this species share features with *Homo erectus* and anatomically with modern *Homo sapiens*. Its brain was nearly as large as that of *Homo sapiens.*

The information provided above is based on mainstream academic knowledge accepted by the majority of the establishment. This knowledge is biased by evolutionary thinking aiming to discover an 'evolutionary tree' showing the gradual development of man. However, the real development of man is much more complex and does not fit the evolutionary model. A more realistic picture is presented by Michael Cremo and Richard Thompson in their book *The hidden history of the human* race[2].

They say that, while most paleoanthropologists believe that *Australopithecus* was a direct human ancestor with a very humanlike body walking erect like modern man, several prominent scientists such as Sir Arthur Keith, Louis and Richard Leakey, Charles Oxnard and others opposed this view. They believed that *Australopithecus* was a more apelike creature than supposed by academia. Although capable of walking on the ground *Australopithecus* was also "at home in the trees, capable of climbing, performing degrees of acrobatics and perhaps of arm suspension." The famous fossil of the 3.5 million year old "Lucy"', found in 1974 in Ethiopia by Donald Johanson, was called *Australopithecus afarensis*. When it was accepted that "Lucy" fitted into the lineage leading to *Homo sapiens* it was assumed that she walked upright in a human fashion ignoring the fact that her shoulders, arms and wrists were adapted for tree climbing.

In 1960 Jonathan Leakey found a skull which had the capacity of about 680 cubic centimeters. It was believed that it belonged to the first true human and was called *Homo habilis*. It was depicted as having a human like body with an apelike head. However, in 1987 a full skeleton was discovered and there was no doubt that *H. habilis* had a small apelike body only about a meter tall, was climbing trees and therefore could not have belonged to the Homo genus.

Cremo writes: "So in the end, we find that *Homo habilis* is about as substantial as a desert mirage, appearing now humanlike, now apelike, now real, now unreal, according to tendency of the viewer."... "This demonstrates once more important characteristic of paleoanthropological evidence – it is often subject to multiple, contradictory interpretations. Partisan considerations often determine which view prevails at any point in time." [3]

The problem with the fossil history of man is that it does not show continuous evolutionary progress from primitive to present-day man, but rather shows the punctuated arising of different groups of hominins in different parts of the world. Fossil records can identify physical changes and even increases in cranial capacity but they do not show the most important aspect of man – the development of his mind and the social and cultural level of hominins living millions of years ago.

Neanderthal man

Neanderthals were the most important hominins prior to the arrival of *Homo sapiens* because they left an excellent record of themselves. We have many fossils of Neanderthal man and there is a significant amount of information not only about his body but about his lifestyle as well. Neanderthals lived across Eurasia, as far north and west as Britain and Spain, through parts of the Middle East, Uzbekistan and up to the Altai Mountains from about 400,000 years ago. The latest Neanderthal fossils were found in Gibraltar and are about 28,000 years old. This means Neanderthals overlapped with man in Europe for several thousand years.

Neanderthals' appearance was similar to ours, though they were shorter and stockier with angled cheekbones. They had prominent brow ridges, shorter limb proportions, a wider, barrel-shaped rib cage, a reduced chin and, perhaps most notably, a large nose, which was much larger in both length and width and placed somewhat higher on the face than in modern humans. Evidence suggests that although they were comparable in height, they were much stronger than modern humans with particularly strong arms and hands. Neanderthal males stood 164 to 168 cm (5'4½"– 5' 6") and females 152 to 156 cm (5' – 5'1½") tall. Their average cranial capacity of 1,600 cubic centimeters was notably larger than the average for modern humans, indicating that their brain size was larger. Neanderthals lived during the Ice Age and their stocky bodies were well adjusted to survive cold winters of Eurasia. They often took shelter from the ice, snow, and otherwise unpleasant weather in plentiful limestone caves. Many of their fossils have been found in these caves leading to the popular term "cave men".

In spite of discovering many Neanderthal fossils and dwelling places there is still a raging controversy concerning his intellectual and cultural capacities. Early researchers presented Neanderthals as very primitive, brutal, grunting carnivores living in caves. Later researchers claim that they could speak, buried their dead and cared for the sick. It is not certain if these new qualities are based on new facts or on new interpretations of old data. Very often new generations of scientists want to distinguish themselves from older colleagues by proposing contrary hypotheses.

Neanderthals were more advanced tool users that other earlier hominin groups. Their stone working skills were impressive but they rarely if ever made tools from other long-lasting materials. They had many highly developed tools and weapons – such as spears for killing mammoths.

No substantial evidence has been found for symbolic behaviors among these hominins, or for the production of symbolic objects - certainly not before contact had been made with modern humans. Even the occasional Neanderthal practice of burying the dead may have been simply a way of discouraging animal raids into their quarters. Neanderthal burials lack the grave artifacts that would attest to ritual and belief in an afterlife. The Neanderthals lacked the spark of creativity that, in the end, distinguished them from *H. sapiens.*

The key question is how Neanderthals differed genetically from modern man. The results of an extensive DNA analysis of a 50,000 year old toe bone belonging to a Neanderthal girl, which was unearthed in a cave in Siberia in 2010, provided very important information. For the first time scientists completely sequenced the fossil's nuclear DNA[4] to the same extent and quality as that of genomes sequenced from present-day people. This amazing research has revealed that there is now conclusive evidence that Neanderthals bred with *Homo sapiens* and as a result about 1.5 to 2.1 percent of the DNA of all people with European ancestry can be traced to Neanderthals. About 35 to 70 percent of the Neanderthal genome persists in the gene pool of people today. Proportions of Neanderthal DNA are higher among Asians and Native Americans.

Results of the analysis show that 87 genes responsible for making proteins in cells are different between modern humans and Neanderthals. Intriguingly, some of the gene differences are ones involved in both immune responses and the development of brain cells in people. Genetic studies discovered that vast numbers of Neanderthal genes are not carried by modern man. This is a strong indication that the genes were harmful to human–Neanderthal hybrids and their descendants and therefore were eliminated.

The results from new studies confirm the Neanderthal's humanity, and show that their genomes and ours are more than 99.8 percent identical, differing by about 3 million bases. In comparison, humans and chimpanzees are about 96 percent similar, and humans and cats are 90 percent alike.

Further genetic study shows that the genome of one of our ancient ancestors, the Denisovans Siberian Neanderthals, contains a segment of DNA that seems to have come from another species that is currently unknown to science, one that is neither human nor Neanderthal.

In spite of so much evidence, Neanderthals are still the subject of several controversies. One important question is where Neanderthals came from. Scientific opinion supports the view that Neanderthals came out of Africa. This belief is based on an assumption that Africa was the origin of the human race, therefore any species related to humans must have come out of this continent. However, the facts do not confirm this hypothesis because no fossils of Neanderthals have been found in Africa.

Looking at the spread of Neanderthals across Eurasia it is possible that Neanderthals could have been linked to *Homo heidelbergensis* who left Africa about 600,000 years ago. Therefore, Neanderthals could have originated somewhere else, like the Middle East, and spread to Europe and Asia. It is estimated that early cousins of Neanderthals – Denisovans lived in the Altai Mountains as early as 400,000 years ago.

Another controversy preoccupying scientists is the cause of Neanderthals extinction. There is a long list of theories such as diseases, malnutrition, genocide, climate change etc. but none of them are very convincing. It is true that the Neanderthals disappeared after *Homo sapiens* arrived in Europe about 36,000 years ago but by this time the Neanderthal population was already drastically reduced to a few thousand and nearly extinct. In fact, new genetic evidence suggests that the population in Europe between 38,000 and 70,000 years ago hovered at an average of 1,500 females of reproductive age. Because of such a low population density interaction between humans and Neanderthals was very limited and the competition for resources had very little effect on the population size. So we have to look for other causes of extinction.

Homo sapiens

There is no doubt that the origins of man are linked with Africa where during the last 5 million years a multitude of different hominin groups existed. Information about them is very scarce, often based on fossils consisting of a few bones or part of a skull, and the dating of these fossils is very unreliable. There is no information on how these groups were related to each other and any proposed evolutionary links are based on the assumption that a more advanced group originated from a less advanced group, which is not confirmed by any scientific data. Many of these groups existed during the same period or had huge lapses of time between them. Some of these groups dwelled for only a few hundred thousand years. The paleontological picture is not very clear and coherent. The only conclusion we can draw is that the road leading to *Homo sapiens* was very obscure.

There is overwhelming scientific evidence showing that *Homo sapiens* suddenly appeared in Africa about 200,000 years ago. The dating of this event is supported by the earliest human fossils from this period and the mitochondrial DNA inherited through the maternal line. Modern people's mitochondrial DNA can all be traced back to the common ancestor, an 'Eve' that lived 200,000 years ago.

Homo sapiens lived in Africa for about 130,000 years and did not show any tendencies to migrate. Suddenly about 70,000 years ago men started cooperating with each other and formed a strongly bonded society. These groups started moving out of Africa not because of overcrowding and a lack of food but because of the inner drive to conquer the world[5]. It is estimated that probably no more than a few hundred people left Africa. They moved in an Easterly direction, probably along the coasts of Arabian Peninsula, India, Burma and Malaysia. On the way they encountered good, fertile lands with plenty of food and many animals to hunt where they would find an easy and comfortable life[6]. However, not knowing where they were going, they still did not stop pushing forward. When they arrived in South East Asia they did not stop there. Instead they built boats and sailed eastwards settling in Australia and the most remote Pacific archipelagos and islands. They also moved in other directions arriving in Central Asia about 45,000 years ago and in Siberia and The Arctic about 35,000 years ago. From The Arctic they crossed to America and travelled to South America.

The situation in Europe could hardly have been more different. Archaeological records show that in this part of the world, after many millennia of almost no progress at all, suddenly human culture seemed to take off like an explosion with the appearance of Cro-Magnon man. They lived in Europe between 40,000 and 10,000 years ago and genetic testing indicates that they came from Central Asia. They are virtually identical to modern man, being tall and muscular and slightly more robust than most modern humans. They were skilled hunters, toolmakers and artists famous for their cave art. They had a high cranium, a broad and upright face, and their cranial capacity was larger than modern humans. The males were as tall as 180 cm (5’11”).

Cro-Magnon people lived in tents and other man-made shelters in large groups. They were nomadic hunter-gatherers and had elaborate rituals for hunting, birth and death. Multiple burials are common in the areas where they were found. Their graves have contained special artifacts suggesting the belief in an afterlife.

They also made tools from bones and antlers such as spear tips and harpoons and had a broad knowledge of the properties of these materials. They invented shafts and handles for their knives, securing their blades with bitumen. Even more significantly, they were the first to leave extensive works of art, such as the cave paintings at Lascaux and Altamira. They kept records on bone and stone plaques; they made music on wind instruments; they crafted elaborate personal adornments; they carved figures of animals and pregnant women. And then about 10,000 years ago they disappeared without a trace.

It is believed that in Europe modern man consisted of a mixture of three groups. One group was the hunter gatherers which arrived from Asia about 40,000 years ago. The next group consisted of wayfarers from the Middle East who introduced modern agriculture about 9,000 to 7,000 years ago, and recently genetic studies discovered a new group joining the other two about 5,000 years ago. This new group of people came from Central Asia and with them they brought bronze age technology.

Comments

The arising of *Homo sapiens* is still a deep mystery. The existence of a large number of various early hominin groups causes further confusion. These groups appeared suddenly and after a short period of time, disappeared without any obvious reasons. Hominin groups were at various stages of development but there is no clear chronological progress, although the latest hominins were more advanced than earlier ones. The biggest problem is that the direct ancestor of *Homo sapiens* has not been identified. It has now been confirmed that our closest and well researched cousins, the Neanderthals were not our progenitors. Academic sources indicate that it was *Homo heidelbergensis* but this is based more on the belief that *Homo sapiens* must have an ancestor than on facts. So it looks like *Homo sapiens* arrived suddenly out of nowhere. The most important aspect of *Homo sapiens* is the development in a very short period of time of an incredibly complex and advanced human brain. The human brain will be discussed in the next chapter.

CHAPTER 9

The human brain and mind

THE BRAIN

Brain structure

There is no question that our large and sophisticated brain gives *Homo sapiens* an extraordinary advantage in the world. Still, the human brain is an incredibly demanding organ, taking up only about 2 percent of the body's mass yet using more than 20 percent of the body's energy. The development of man discussed in the previous chapter included the development of the brain. However, what we can learn from fossil records shows only the cranial capacity of the skull, in other words the amount of space taken up by the brain. It does not tell us about the intellectual capacity and the intelligence of the person. We know that until about 2 million years ago none of our ancestors had a brain larger than an ape's when compared to their body size. From then we can see the continuous increase of cranial capacity culminating with Neanderthal's brain having an average volume of about 1600 cubic centimeters. The brain of modern man is surprisingly smaller, having a capacity ranging from about 1000 cc up to about 1500 cc. Anatomy of the brain is well covered in scientific and popular literature therefore I will only look into its more interesting features.

The human brain is an extremely complex organ. It is composed of neurons*, glial cells* and blood vessels. An adult human brain is estimated to contain about 86 billion neurons, with roughly an equal number of non-neuronal cells. Out of these, 19 percent of all brain neurons are located in the cerebral cortex* and 80 percent in the cerebellum*. If we add to this list at least 100 trillion synapses* linking the neurons, we are talking about a gigantic network the complexity of which is beyond human imagination. These cells must be connected in a well organized network to be able to perform so many different functions. The brain is not a bag of neurons packed together randomly, but a precise instrument operating with high accuracy and reliability. The truth is we haven't the faintest idea how this network is constructed and how it operates. Our knowledge ends with individual neurons, synapses and with identifying the parts of the brain that perform certain functions. These parts are called Brodmann areas and were originally defined by the German anatomist Korbinian Brodmann who, in 1909, published his maps of 52 cortical* areas in humans and some animals. To this day his maps are still the main source of our information about the brain, showing how little progress has been made in over a century. This lack of progress can only be attributed to the enormous complexity of this network.

One of the most important morphological features of the brain are the blood-brain barriers which separate circulating blood from the brain's extracellular fluid. The blood–brain barrier allows the passage of water, some gases, and lipid-soluble molecules by passive diffusion, as well as the selective transport of molecules such as glucose and amino acids that are crucial to neural functions. The blood–brain barrier acts very effectively to protect the brain from many chemicals and most bacteria. Thus, infections of the brain are very rare, however when they do occur they are often very serious as antibodies* are too large to cross this barrier.

Brain drainage system

One of the most recent and intriguing discoveries made is the brain waste drainage system[1]. In the rest of the body a network of special vessels carry lymph*, a fluid that removes excess plasma, dead blood cells, debris and other waste. But the brain is different. Instead of lymph, the brain is bathed in cerebrospinal fluid*. The brain generates something like 1.5 kilograms of waste products a year, equivalent to its own weight. This waste consists of the products of cell metabolism, broken proteins, non-functioning cell complex molecules and some toxins therefore it is vital that this waste is removed as quickly as possible.

It has been discovered that the drainage system consists of a layer of pipes that surround the brain's existing blood vessels. It serves the brain in much the same way as the lymph system serves the rest of the body by draining away waste products. Star shaped glial cells called astrocytes*, which have several long tentacles ending in wide "feet", form a network of conduits around the outside of blood vessels inside the brain. These "feet" completely surround the arteries, capillaries and veins and form what looks like pipe cladding with a cavity between the vessels and the "feet". Fluid containing waste flows in this cavity propelled by the pulsing flow of blood. Cavities around small veins merge into cavities around larger veins and eventually join the lymphatic system in the neck. Waste from the lymphatic system enters the blood vessels and is processed by the liver or filtered by the kidneys. This research also found that most of the waste removal takes place during sleep when waste is cleared twice as fast as when awake.

From an engineering point of view this is a very ingenious solution using an existing system of blood vessels to support the draining system and to use pulsing blood to pump fluid around the brain.

The functions of the brain

You may have heard of the BRAIN[2] Initiative, a grandiose plan to "map the human brain" initiated in 2013 by President Obama who wanted to be remembered as a great scientific visionary. However this project has nothing to do with science. The reality is that we are unable to map the brains of even the simplest living organisms. For example, scientists are still unable to explain the workings of a simple network consisting of a few thousand neurons belonging to a Hydra, a small fresh water animal.[3] This example shows the extent of our knowledge and the problems facing investigators mapping 86 billion neurons in the human brain.

So what do we know about the brain? The latest technology used to study the brain is functional Magnetic Resonance Imaging* (fMRI). Some researchers allege that fMRI can even 'read' human thoughts and analyze human behavior. There is a lot of hype and unsubstantiated claims related to this technology. What fMRI actually does measure is the amount of oxygen in the blood flowing to the part of the brain being tested. It is known that when neurons are firing they use up much more oxygen than when at rest. However, the data obtained from fMRI tests has nothing to do with specific signals in the brain and is not able to provide any information on how the brain works.

The smallest volume of brain which can be analyzed is about 1 cubic millimeter. A typical volume used for tests contains a few million neurons and tens of billions of synapses. Effectively this technology provides a more detailed brain map, but more importantly, it provides information about the dynamic responses of the brain.

To help us gain some insight into how the brain operates we need to look at some brain sensing functions such as vision and hearing. These functions are easier to test because we can stimulate the brain with a controlled range of different inputs and monitor the brain's responses with electrodes.

Vision

The part of the brain which is involved with vision is called the visual cortex*. This is the most studied part of the brain, however much of the information was obtained by looking at the brains of macaque monkeys[4]. The visual cortex is placed right at the back of the head and consists of six areas. Anatomy of the cortex is well known, however it is too complex to outline here hence we will try to show what is known about optical signal processing.

The most important and complex part of the eye is the retina. The function of the retina is to transform an optical image into a series of electrical pulses which are sent to the visual cortex. Light falling on the retina is detected by photoreceptors including rods which are sensitive to very weak light, and cones which sense color. Signals from photoreceptors are passed through three layers of cells which perform complex signal processing functions. The third layer consisting of ganglion* cells generates special signals to be transmitted to the brain. The ganglion signals consist of a series of electrical impulses or 'spikes' having a relatively constant amplitude but variable frequency. These signals are fed to the optic nerve which takes them to the brain. However, before signals are sent to the brain, the first image processing takes place in the retina where about 130 million photoreceptors detect light. From there only about 1.2 million ganglion cells transmit information from the retina to the brain. So the signal must be processed in such a clever way as to not lose any image information. The transmitted signal not only contains information about the image color and intensity, but also edge detection and movement.

The retina is divided into two parts and each part of the retina has its own optical nerve. Signals from both eyes do not go straight to the visual cortex but follow a convoluted path. Optic nerves from each eye connect to a junction box called the optic chiasm*.

From the optic chiasm 90 percent of the signal is transmitted via the optic tract to the lateral geniculate nucleus*, or LGN, in the thalamus* and the rest to other parts in the mid-brain. The neurons of the LGN then relay the visual image to the primary visual cortex. The LGN is not just a simple relay station but is also another centre for processing. The visual cortex has a map of the retina, meaning that any point in the retina corresponds to a unique point in the visual cortex. Signal processing in the visual cortex is still not fully understood.

The real miracle of the visual system is that the brain, which receives a series of electrical impulses from the retina via a million parallel optic nerve fibers, generates a virtual 'image' of the original picture in the cortex. This image corresponds in great detail to the real object. It is not distorted, it has true colors and it has 3 dimensions. We can judge the object distance from us and we can detect its movement. What is even more remarkable is that the brain can generate images without receiving any optical signals. When we dream we sometimes perceive very detailed images not related to reality, and hallucinogenic drugs can induce bizarre images.

This brief description should give us an idea of the immense complexity of the visual system. This system could not have evolved in small steps but the eye, retina and brain processing had to be implemented at the same time.

Hearing

Studying the human auditory system has led to the discovery of another amazing function of the brain. The auditory signal is generated in the ear which has a very ingenuous design. The key component of the ear is the cochlea which is responsible for changing acoustic waves into electrical signals to be processed by the brain. The cochlea is a hollow cavity of bone, spiraled like a snail shell. This fluid filled cavity is divided into two longitudinal chambers separated by a partition. The organ of Corti* is the sensory organ of hearing, which is situated along the fluid separating partition chambers in the tapered tube of the cochlea. It is responsible for the transduction of auditory signals into electrical signals which consist of electrical impulses similar to the spikes in the visual system. Transduction occurs through the vibrations of structures in the inner ear causing a wave motion in the cochlea's fluid from the base to the top of the spiral. The moving liquid causes the movement of hair cells at the organ of Corti which produces electrical signals. The cochlea has over 32,000 hair cells which detect the motion of those waves and excite the neurons of the auditory nerve. The wide end of the cochlea, where sounds enter from the middle ear, encodes the higher end of the audible frequency range while the top end of the cochlea encodes the lower end of the frequency range. The cochlea is a marvel of physiological engineering and acts as both a frequency analyzer and nonlinear acoustic amplifier.

Electrical signals from the cochlea are sent to the auditory cortex in the brain via four stages, the functions of which are still to be identified. One of the most important stages is the medial geniculate nucleus*, or MGN, which is part of the auditory thalamus. The MGN is involved in auditory processing and in directing one's attention towards specific auditory stimuli. In fact we know very little of how this part of the brain works. Signals from the MGN are sent to the auditory cortex which is placed on both sides of the brain.

Again we know very little about auditory signal processing because it is so complex. We do know that the auditory system provides information about sound intensity, sound pitch (frequency), sound location, sound duration and binaural reception. This system is somehow linked to the visual processing system.

Looking at these two relatively accessible and easy to follow brain processing systems we face the same problem. The signal processing is beyond our comprehension and although we can identify the signal transmission paths and follow signals along these routes we are still unable to make any sense of it.

The brain positioning system

Recent research shows that mammal's brains, including that of man[5], have a very sophisticated mapping and positioning system[6] which can be compared to a performance of Global Positioning System (GPS). When an animal moves, neurons generate a map in a part of the brain called the hippocampus* which helps it to find a location and enables it to return to its original position. The neurons responsible for this mapping are called grid cells. The grid cells identify the location of an animal using a hexagonal grid. When the animal is in a specific grid position a corresponding neuron is fired. When the animal moves to another position about 30 cm away, another neuron is fired. This way the animal knows its location without referring to any external clues.

The positioning mechanism is very complex and includes 'speed cells' which fire at different speeds acting as a speedometer. There are also direction cells which fire when the animal faces a specific direction and there are 'border cells' that provide information when an animal is nearing the border of its local environment.

We do not know much about this positioning system, but what we know indicates that the system is highly complex and uses very sophisticated signal processing.

Genetic development of the brain

We know that almost every cell in our body contains exactly the same set of genes, but inside individual cells some genes are active whilst others are not. Only when the genes are switched on they are capable of producing proteins or enzymes. This process is called gene switching and was discussed in chapter 6. We know that at least a third of approximately 20,000 different genes that make up the human genome are primarily active in the brain. This means that the brain has the most complex structure in the whole body.

Scientists have identified clusters of genes in the brain that could determine human intelligence. There are two clusters of hundreds of individual genes that are thought to influence all our cognitive functions - including memory, attention, processing speed and reasoning, but we do not know how.

A significant insight into the development of the brain was provided by Bruce Lahn and his team at the Howard Hughes Medical Institute at the University of Chicago. It is worth reporting his findings as they help to explain improvements in the human brain[7].

Lahn investigated two genes, microcephalin and ASPM[8] across different human populations. He assumed that these genes played a role in our cerebral development since the mutation of either of these genes leads to the severe condition called microcephaly - a clinical syndrome where the brain develops to a much smaller size than normal[9]. Both genes are known to regulate brain size, and related studies suggest they might control cell growth in the developing brain.

By sequencing human DNA samples the Lahn group found that for both the microcephalin and ASPM genes, one predominant variant exists. Using mutation rates as a kind of "molecular clock", the team determined that the prevalent microcephalin variant emerged approximately 37,000 years ago, while the dominant ASPM variant appeared about 5,800 years ago. By sequencing the versions of these genes carried by 1,200 people from across the globe they found that the preferred microcephalin variant was common in all but sub-Saharan Africa, while the preferred ASPM variant was common only in Europeans and Middle-Easterners[10].

The emergence of the microcephalin variant coincides with archaeological estimates of the movement of Cro-Magnons into Europe about 40,000 years ago and the development of more sophisticated society, rituals, beliefs and art. The appearance of the ASPM variant coincides with the development of the Sumerian, Indus Valley and Egyptian civilizations around 5,500 years ago.

Bruce Lahn said that: "People in many fields, including evolutionary biology, anthropology and sociology, have long debated whether the evolution of the human brain was a special event, ... I believe that our study settles this question by showing that it was."

One of the study's major surprises is the relatively large number of genes that have contributed to human brain development. They came to the conclusion that the development of the human brain could be the result of thousands of mutations in perhaps hundreds or thousands of genes—and even that is a conservative estimate. They believed that it was nothing short of spectacular that so many mutations in so many genes were acquired during the mere 20-25 million years of time.

Lahn's research confirms that the human brain developed in large steps and much faster than could be accounted for by normal mutation rates. He believes that this was the result of accelerated evolution, but he is unable to explain how this acceleration happened and what the mechanism behind it was.

His research supports the hypothesis that the human genome was manipulated and changed several times since the arising of *Homo sapiens,* and the last time it happened was only about 6,000 years ago.

Several other scientists also observed more rapid alterations of the human genome than would ordinarily have resulted from random mutations. These unexplained changes were estimated to be a hundred times greater than known mutation rates. The mechanisms of mutations are well researched and known, therefore they could not account for these high rates. As such it is conclusive that other mechanisms must have taken place to change these genes at this alarmingly accelerated rate. Since gene changes are well proven, this increase in mutation rates is a visible and tangible result of artificial gene modification.

THE MIND

The human mind has many incredible abilities and features, some of which we are not always aware of. The ability to communicate using language, the ability to learn, intuition guiding us through complex situations, logical and abstract thinking, inventiveness, the drive to discover new things, and appreciation of art are just a few examples from a long list.

While knowledge of the workings of our brain is very limited, the opposite applies to the human mind where our knowledge is much broader. The amount of information on the human mind is staggering therefore I will concentrate on just a few aspects of the mind which makes us so different from primates.

Consciousness

Our consciousness is a fundamental aspect of our existence. To quote the philosopher David Chalmers: "There's nothing we know about more directly.... but at the same time it's the most mysterious phenomenon in the universe".

There are not many issues which divide scientists and philosophers more than consciousness. Consciousness is very difficult to describe. It is one of those nebulous phenomena which everybody has their own definition of, or rather everybody experiences but cannot explain. One description I would add is that consciousness is something that enables man to leave his body and look at himself from the outside and be aware of himself as a separate entity.

Does everyone have the same level of consciousness? The politically correct answer is yes, but observations tell us otherwise. There is a wide spectrum of consciousness[11] and people can be at any level of this spectrum. Our level of consciousness determines our social responsibilities and interactions. One could say that people on a low level of consciousness care only about themselves and their closest family. People on higher levels care about strangers, society, the environment, etc, and take part in charitable activities. So our consciousness defines our position and contribution to the world we live in.

The level of consciousness is linked with our abilities to be introspective which involves turning our thoughts inwards and reflecting upon our behavior. Scientists have observed noticeable variations in people's abilities to introspect and therefore their level of consciousness.

A team of researchers, led by Prof. Geraint Rees from University College London, suggested that the volume of gray matter in the anterior prefrontal cortex of the brain, which lies right behind our eyes, is a strong indicator of a person's introspective ability. This means that the physical structure of the brain could influence our level of consciousness.

Have people always possessed consciousness? The world of science believes that man has had an awareness of himself right from the very beginning of his existence, at least for tens of thousands of years. However, Julian Jaynes (1920-1997), a professor of psychology from Princeton University, disagreed. In 1976 he advanced a very controversial hypothesis[12] that consciousness did not arise at the start of human evolution, but was only formed in the last 3,000 years. Primitive man had, as he called it, a bicameral mind in which both hemispheres did not cooperate with each other in the way they do in conscious man. Bicameral people were not aware of the existence of their own thoughts and internal dialogues, only experiencing them as a kind of auditory hallucination or voices coming from the outside directing their actions.

Jaynes based his hypothesis on the studies of the Sumerian civilization and on early Greek literature, particularly the works of Homer. He considered that the theocratic Sumerian civilization with its rigid hierarchy, resembling a bee hive where everyone had their place fixed in advance, reflected the bicameral mind. Ordinary Sumerians did not possess full consciousness because they blindly carried out orders from the gods.

According to Jaynes, the development of consciousness took place in Mesopotamia and in Greece as recently as 1,000 BC, and in other places even later. However he is not able, in a convincing way, to explain the mechanism of the arising of consciousness which took place in such a short time, even though essential changes had to occur in the functioning of the brain. But recent genetic research could throw some light on this. We now know that the genetic makeup of the brain was modified about 6,000 years ago. It is only after these modifications were established in the wider population that we are able to observe the development of consciousness.

Abstract thinking

What distinguishes man from animals is his ability to think in an abstract way. The development of abstract thinking happened during the first civilizations, about 5,000 years ago, when social abstract concepts such as justice and freedom were used for the first time[13].

The ability of abstract thinking provided man with exceptional tools to develop the sciences and technology and opened a world far beyond his immediate perception. It enabled man to obtain the proof of certain phenomena without the necessity for obtaining material objects. These abilities include spatial reasoning and mentally manipulating and rotating objects.

Abstract thinking enabled man to generate and understand art. Before art is generated it first exists in an abstract form in the artist's mind. Artwork that reshapes the natural world for expressive purposes is called abstract. Art now plays a very important role in representing complex feelings, emotions and creating beautiful objects.

Language also uses some abstract components such as metaphors, symbols and analogies. Abstract thinking helps to understanding the relationships between verbal and non-verbal ideas. Poetry is another linguistic form which is a product of the abstract process.

Music is one of the most difficult forms of abstract art to comprehend. Its emotional reception depends on our understanding of the abstract meaning of tones. The gift to create music is possessed by very few people. Famous composers such as Mozart or Beethoven said that they had all their compositions ready in their head to be written down. They believed that it was a gift from God.

Another kind of abstract thinking is used in mathematics where any dependence on real world objects is removed and the use of generalization has wider applications. Techniques and methods used in one area can be applied to prove results in another related area. Another example of abstract tools is logic which helps to prove phenomena by abstract reasoning. It is abstract thinking which has enabled man to move to a higher level of existence and to develop a very sophisticated world.

Comments

The human brain is the highest achievement of biological design. It is so amazingly complex and different from anything we have developed ourselves. Why the brain needs so many neurons and connections is and will remain a mystery. It is not a computer, although it uses digital types of signals. The propagation of its digital pulses is not very fast but the brain's operating speed is enhanced by parallel signal processing, for example, a million optical fibers carry signals from the eyes. Our brain is not only an electronic device but it also utilizes chemical signals conveyed by enzymes and hormones. Its connections could be compared to wires, but signals passing through synapses are controlled by many other factors. It is a dynamic structure which can change its wiring connections and consequently evades our efforts to get a glimpse of it.

The complexity of the human brain explains why the development of *Homo sapiens* took so long and required so many iterations. Although the mammal's brain looks very similar it was designed to control relatively simple and repetitive processes and therefore have limited flexibility. Animal behavior is either coded in their genes or is learnt by example, imitation and experience. The human brain, besides these facilities, has an enormous range of different properties which enable man to have consciousness, abstract thinking and analytical analysis abilities, enormous learning capacity and a large memory. So the operating programmed of the human brain is quite different from that of animals. But this programmed offers man the facilities to develop higher consciousness and to improve his learning and reasoning. So the human brain can change by altering internal connections. It is well proven that in cases where a part of the brain has been damaged, another part can take over the same functions.

A particular property of the human brain which could not have arisen as a result of simple genetic modifications is its ability for further development. This development is most visible in the domain of consciousness. It is known that the genetic makeup of the human brain has not changed for the last 6,000 years, and yet we observe significant developments in human thinking and consciousness such as our concern for the environment, climate change and the Earth's future. This shows that the human brain has a built in function enabling further development which is not based on genetic or evolutionary mechanisms. Therefore one could predict that the development of the human race will result in a more humanitarian society.

So will the functioning of the brain eventually be discovered? This is a more philosophical than technical question. Is it possible that a system of certain complexity could be able to comprehend itself? I believe that to understand a complex system one needs a tool which is much more sophisticated or advanced than the system being investigated. For example, in engineering to analyze a complex digital system we need an instrument which is even more advanced. One could observe from real life interactions, where a less intelligent person would not be able to understand a more intelligent person, or a man with a low level of consciousness would not understand a man on a higher level of consciousness, while the opposite is true.

Since our brain has been designed by beings more intelligent than us, they would not include in our brain faculty the function to comprehend itself as there would be no need for it. I believe that the brain's operating system is based on completely new principles which are beyond our comprehension.

The hypothesis of intelligent design of the brain is confirmed by human genome research. This research shows that genetic changes could not have resulted from normal mutation rates because they were about a hundred times faster. The human brain was improved at least twice during its history. Its development was only completed about 6,000 years ago when it was significantly enhanced resulting in a new conscious man who built the first civilizations. This was the beginning of a new era in the progress of mankind.

To summarize, you the reader, may not learn much about the brain from these few pages, but even if you undertook extensive and in-depth research into this subject, your quest for knowledge would still not be fully satisfied. The fact is that we know more about the functioning of distant galaxies than we do about our own brains.

The opposite is true regarding our mind, where huge encyclopedic knowledge has been accumulated due to extensive testing. This information helps us to understand ourselves better and therefore improves the quality of our lives.

CHAPTER 10

Evolution

Understanding Darwin's evolution

It is widely believed that Darwin's evolution explains the origins and development of everything in the biological domain. This expression is used so frequently without any hesitation because it provides answers to every question, solutions to every problem and explanations of any new discovery or paradigm.

Michael Behe writes: "In recent years Darwin's intellectual descendants have been aggressively pushing their idea on the public as a sort of biological theory-of-everything. ... The penchant for viewing the world through Darwinian glasses has spilled over into the humanities, law and politics. Because of the rhetorical fog that surrounds discussion of evolution, it's hard for the public to decide what is solid and what is illusory."[1]

One would suspect that people use the word "evolution" automatically and do not think about its correctness or applicability. It must be right because everybody in the academic world is using it. It sounds "scientific" and is politically correct. No one will be offended if it's applied to race or gender issues. It is atheistic so no one can be accused of religious bias or supernatural or paranormal inclinations. This attitude very much narrows our range of options and makes us lazy, because it does not force us to look for other scientific answers. It doesn't matter how implausible our explanation is, we are always safe with "evolution". We feel secure when using this word because effectively it is someone else's responsibility to provide the explanation of how evolution could have arranged the development of this function or that structure.

When reading scientific literature describing the development of life we find anthropomorphic attributes of evolution. Writers pronounce that evolution anticipated, predicted, decided, choose the right solutions, has intuition, is dormant or accelerates, acts on the selected sample, guesses which properties are needed, knows how to select exactly the right compounds from millions of chemicals, and can design the most advanced nano-structures which are beyond human understanding. It can design complex control systems, coding and decoding systems as good as the most sophisticated electronic system. It is very clever, a genius, a god because it replaces God.

It is worrying that nobody provides an explanation of how all this could have been accomplished. I do not expect experiments to prove the evolutionary hypothesis, but at least attempts to provide some theoretical guesses, basic calculations and logical structures. One cannot find any flow charts showing how evolution could have progressed along a certain path, one cannot find any calculations of probability of the arising of certain structures through random mutations. Darwinian thinking is so entrenched that evolutionists say that "we do not have to prove evolution", although evolution is still a hypothesis. Franklin Harold, professor of biochemistry at Colorado State University summarized this neatly by stating: "Acceptance of the Darwinian thesis rests on its general plausibility, on the absence of positive evidence to the contrary, and on the lack of a credible alternative."[2] However he does not mention any credible scientific evidence supporting evolution. A literature search reveals many thousands of papers describing structures and functions of biological organisms but only very few attempt to look at how such structures could have evolved and do not provide any details beyond general mandatory statements.

Some scientists find that evolution is not able to explain certain molecular developments and therefore assign imaginary functions to the evolutionary process. One of the most common mistakes is an assumption that the living cell can select, from the environment, a suitable material needed for its development and somehow pass this information to the genes. For example, in chapter 5, a catalyst helping to split water was described. This catalyst works with 3 large proteins with structures that bond to calcium and manganese atoms. Evolutionists state that since this calcium-manganese structure could have existed in nature, the cell adopted it as a catalyst. To implement it, the cell would had to have known in advance that this was the right compound and tell the DNA to code 3 proteins which would work with these metals. This means that the DNA would have to receive information about the choice of metals, which is against the central dogma[3] of evolution.

From the three key elements of evolution: random mutations, natural selection and common descent, only common descent is not contentious. The results of genetic studies prove, beyond any doubt, the existence of common ancestry of all living creatures on Earth. It is widely known that the human genome shares 50 percent of its DNA with bananas, but not many people know that we share, on average, 6.3 percent of our proteins with simple organisms such as bacteria.

Most scientists and the general public support the hypothesis of evolution because they are familiar with the gradual development of the animal world. This progress is easy to grasp even for a layman, but the process of this development has never been properly explained.

Behe states, "On the surface, Darwin's theory of evolution is seductively simple and unlike many theories in physics or chemistry, can be summarized succinctly with no math. ... In order to forge the many complex structures of life, a Darwinian process would have to take numerous coherent steps, a series of beneficial mutations that successively build on each other, leading to a complex outcome." [4] However there is no proof that evolution is able to deliver a large number of advantageous mutations.

Since there are so many misconceptions about evolution, and sometimes pseudo-evolutionary explanations are based on sheer ignorance, it is necessary to provide a brief resume of the foundation of evolution. We will look into two pillars of evolution: random mutation and natural selection.

Mutations

We are aware of the negative associations of gene mutations because they have, in our minds, been linked with severe diseases including some cancers. The devastating consequences of thalidomide, prescribed for morning sickness in pregnant women in the 1950-60's, still reverberates in our memories.

A mutation is a permanent alteration of the nucleotide sequence in DNA which results in protein coding errors. Mutations can be caused by external or internal factors, or they may be caused by errors in the cellular machinery. Physical or chemical agents that induce mutations in DNA are called mutagens.

External mutagens include ultraviolet, especially UVB radiation from the Sun, and other radiation frequencies, including x-rays, gamma rays and alpha particles, high temperature, certain plant toxins, viruses and man-made mutagenic chemicals. Internal factors include errors during DNA replication which can lead to genetic changes, and toxic by-products of cellular metabolism.

Most mutations have a destructive effect and tests have shown that out of all stomatitis virus gene changes, 39.6% were lethal, 31.2% were non-lethal deleterious, 27.1% were neutral and only 2.1% were slightly beneficial.[5] Under these circumstances, life would need mechanisms to protect organisms against the effects of mutations.

As we know, Earth has two built in protection mechanisms against harmful solar radiation. Protection against high energy particles coming mainly from the Sun is provided by the Earth's magnetic field as was described in chapter 1. Protection against the Sun's ultraviolet radiation is provided by the ozone layer which is formed in the upper part of the Earth's atmosphere and absorbs about 95 percent of harmful emissions.

Another clever design solution is that the protein coding system can tolerate certain changes to the protein structure without affecting its function. These are known as neutral mutations which constitute about 30% of all mutations. In practice about 70% of mutated organisms are removed from the population and 30% have their DNA coding altered. This is known as DNA drift.

These percentages look bad, but due to cell repair mechanisms mutations do not happen that often and for human DNA the mutation rate is about one in a hundred million per base per generation.[6] It has been observed that the protein coding drift rate in animals is about one amino acid change per 4 million years. However, this drift does not mean that the modified proteins provide improved performance.

Such a low mutation rate makes the process of evolution very difficult. To ensure that beneficial mutations become fixed, the population must be very large. This is illustrated by the following example.

Let us assume that we start with 10 trillion cells (10^{10}), a mutation rate of one in one hundred million, and a beneficial mutation rate of 0.02. Out of these 10 trillion cells only two will gain one beneficial mutation. These two cells begin to divide and we have to wait 33 generations until the population with one beneficial mutation reaches about 10^{10} cells. As a result of the mutation of these cells, two of them will receive one additional beneficial mutation, therefore they will have two beneficial mutations out of a total of 10^{20} cells. For two cells with 3 beneficial mutations we would have to wait 66 generations for the population to reach 10^{30} cells.

This illustration is very simplified and does not identify any specific genes and coding to be changed. It merely illustrates that to accumulate any 3 beneficial mutations, not even necessarily affecting the same function, the population must be very large to start with.

With a smaller population, the fixing of two or more beneficial mutations becomes very difficult as fewer cells will be mutated. Even when a cell receives a beneficial mutation the chances of a second mutation is extremely low, and it is more likely that the next mutation will be lethal or deleterious rather than beneficial.

So in practice we can observe one or two beneficial mutations only in very large populations of organisms such as bacteria and parasites. For example, the resistance of bacteria to antibiotics results from one or two mutations. However, as a result of these limitations any significant changes of animals and plants via the mutation route are practically impossible.

Do mutations drive evolution?

Do we have any proof that mutations improve organisms? There is a huge body of evidence that shows mutations cause harm to organisms. Now we know that the structure of the cell is so complex and its workings must be so precise that any, even the smallest changes could upset its functioning and would cause its obliteration. For example, it has been identified that more than 150 genetic diseases are caused by mutations of the molecular complexes in the electron transport system alone, which are responsible for energy generation in mitochondria. Mutations of the spliceosome which cause errors in splicing genes result in serious genetic diseases in humans such as spinal muscular atrophy and lymphocytic leukemia.

The limits of evolution were discussed by Behe in his book *The Edge of Evolution*. He clarified the most confusing issue about evolution, namely that evolution works but its functionality is very limited. Behe estimated that random mutations could change no more than two bases in one gene. This estimate is based on the development of resistance of bacteria and malaria parasites to drugs.[7]

However, in higher organisms with smaller populations than malaria parasites, the odds of even two mutations in the same gene are extremely low. Behe made the calculations and came to the conclusion that: "No mutation that is of the same complexity as chloroquine resistance in malaria arose by Darwinian evolution in the line leading to humans in the past ten million years."[8]

If we take into consideration that the necessary genetic changes leading from the ape's common ancestors to *Homo sapiens* would have required many thousands of very specific mutations, the whole process of evolving more complex organisms is totally unrealistic.

Since it is impossible to prevent DNA mutations, the integrity of the cell is protected by not having critical information stored in the DNA at all, thus protecting it from mutations. As was discussed in chapter 4, the most important information about cell architecture and its functioning is an innate part of the cell itself and is transmitted from cell to cell during reproduction[9]. It is in this way that the continuation of life and the uniformity of all cells in organisms is guaranteed.

However, a lot of important information is still coded in DNA and this information must also be protected. To do this, cells use very sophisticated repair mechanisms. Cells have complex signaling networks that carefully monitor the integrity of the genome during DNA replication, and to initiate the repair of faulty DNA sequences. Since there are two copies of genes in each cell, when one copy is mutated, the repair mechanism replaces the mutated part using the other gene as a template. If the cell did not have all these protective measures, its DNA mutation rate would be similar to the mutation rate of viruses, which could be up to ten thousand times higher than in animals. This shows that organisms have a very effective system of protection against the consequences of mutations. So, the obvious question arises: Why would evolution provide such extensive protection from the effects of mutations if these mutations were supposedly beneficial to organisms?

Can mutations make new proteins?

Whilst evolutionists propose that random mutations are the source of new proteins they do not provide any calculations of how many such mutations would be needed. To have a better understanding of the sort of numbers we are talking about, here is a simple exercise to calculate the number of random mutation steps required to obtain a code for a small protein having 100 amino acids in its chain. It is worth noting that an average yeast protein consists of 466 amino acids and the largest animal protein uses about 27,000 amino acids. The sequence of amino acids in the chain must be correctly selected otherwise the protein will not perform its function. There is a possibility that two slightly different proteins could perform the same function but this would not change the results of our calculation.

Since proteins are built from 20 amino acids, 20 steps are needed to select the one correct amino acid using the random selection process. To select two correct amino acids $20 \times 20 = 20^2$ steps are required. For three amino acids we need 20^3 steps and so on. Therefore, to obtain the code for 100 amino acids in the correct position we would need 20^{100} steps. This means that in practice we would have to make 20^{100} different DNA strands to be 100 percent sure that we have the code for the correct protein. It is difficult to comprehend how large this number is. It is one with 130 zeros. If we assume that the simplest cell has a few thousand proteins working with each other, the hypothesis that they could arise via random mutations is unsupportable.

Genetic variation

Although mutations could be responsible for the generation of new DNA sequences, their role is very limited since mutation rates are very low and most mutations are harmful. What can be observed in real life is an incredible variety of body shapes and colors of the same species. For example, each person out of 7 billion people living on Earth is different, including identical twins. It is genetic variation which makes us all unique, whether in terms of hair color, eye color or even the shape of our bodies.

This variability is due to an immense pool of genes describing different traits in the population. During sexual reproduction, processes take place which result in a complete mix of the father's and mother's genes, therefore children in general do not look like copies of their parents. During this process, some inherited genes may or may not be activated. This is a complex process, the description of which is beyond the scope of this book, but the result is that animals and plants have a very effective genetic mechanism which helps them to adapt to the changing environment.

This mechanism has been hijacked by evolutionists who use it as proof of evolution. But, in reality, this process has nothing to do with evolution because it does not make completely new genes, it is not responsible for new body parts and most importantly it is not able to generate new species. This process uses existing DNA sequences and shuffles them in such a way that the progeny will have different characteristics. As a result of natural selection, these characteristics will become established in the population.

Popular examples of evolution

Almost everyone is familiar with examples of evolution which are a part of our culture and shape our understanding of this theory. One of the most publicized examples is the change of color of a peppered moth. Initially, the moth was white with black spots and as a result of pollution during the industrial revolution it became black or much darker. Sewall Wright - a famous American geneticist described it as "the clearest case in which a conspicuous evolutionary process has actually been observed".

Before 1811 the black moth variety was extremely rare but by 1850 it was well established and by 1895, dominated the field. The change of color in such a short period of time was not a result of mutations but simply originated from gene variability. It is known that the black variety existed before the industrial revolution but was uncommon because its camouflage did not protect it against predators. When dark camouflage was needed, the moth changed its color as a result of selecting a gene responsible for dark pigment. This gene spread as a result of natural selection, viz. dark color provided good protection against predators. This example does not prove that evolution could generate new body plans or species. It simply confirms the existence of a built in mechanism for rapid adjustment to the changing environment.

The existence of so many breeds of dogs is a good example of genetic variation in action. There are more than 350 breeds, some of them spectacularly different such as Chihuahuas and Great Danes. There are attempts to call this phenomenon 'man-made evolution'. Again, this example has nothing to do with evolution because all these breeds belong to the same species. Breeders simply selected dogs with the required characteristics and rejected dogs with unwanted characteristics, performing 'natural selection'. This process was discovered by Mendel almost 150 years ago, by breeding selected plants. Again, these breeds do not prove evolution because they do not generate new genes.

There are, in school books and popular literature, more examples of evolution such as the giraffe's neck, Darwin's finches, Australian cane toads, etc., but they do not prove evolution because the described changes could have resulted from the gene variability mechanism. We could therefore summarize that there are no proofs showing genuine evolutionary development resulting from random mutations.

Natural selection in small steps

One of the tenets of evolution is that it progresses in small steps via natural selection. What does this mean? When an organism is changed by mutation, this improvement can be fixed in the population if the mutated organism has a higher reproductive rate than the non-mutated organisms of the same species. So, the improvement must be translated into a larger number of offspring. However, this argument has a few weak points. There could be improvements which might not necessarily affect the reproductive rate. For example, the development of some artistic or mathematical abilities would not affect a farmer's food production or number of children.

The rate of reproduction depends on many environmental parameters such as the amount of available food. However, food supply depends on many factors beyond the control of animals such as climate fluctuations, diseases, competition, etc. These factors could have a stronger effect on the number of offspring than some small genetic improvements. Effectively dominant environmental factors act as a camouflage masking any genetic improvements resulting from small beneficial mutations which do not become fixed in the population.

The environmental effect is even more applicable to plants whose reproductive rates depend on the available supply of energy and materials. In the case where plants produce many thousands of seeds, the reproductive rate is dominated by the environment. Being in the right place at the right time is more important for the seed to succeed than any small genetic improvement of the adult plant which could result in producing, say, 5 percent more seeds.

As mutations are accidental, each part of an organism's body has an equal chance of being affected. It is possible that the performance of several parts could be changed simultaneously. So, where population numbers are small, one mutation could improve fitness and procreative results and another mutation could be detrimental, therefore the improvement is not passed on to the offspring.

It can be observed that certain features of animals, such as limb or body size, could change in small increments, but this may not always be caused by DNA mutations. The question is whether such a process is possible at a molecular level. It is difficult to envisage how a small improvement of one molecule could affect the reproductive rate of the cell. These molecules are part of a long chain of reactions and changing only one reaction would not affect the efficiency of the cell's metabolism. This could apply to such metabolic functions as the Krebbs or Calvin cycles, where the improvement of part of the cycle would have no effect on fitness.

In general, it is possible that an increase in fitness could only result from the joint operations of several parts or functions working together, while the modification of each separate part individually would have no effect. In this case, how could these individual improvements evolve?

One parameter, which is frequently used to support the theory of evolution, is time. Evolutionists have managed to persuade the general public to believe that if the process of evolution is long enough anything could happen. It is not easy for the average person to estimate what biological structural changes are possible during say, 10 million years. Therefore evolutionists frequently use the argument, "this was such a long period of time that the evolution of this component was possible". However this argument is very deceptive because what actually matters is not time but population size. To estimate how quickly evolutionary changes could take place, the mutation rate must be multiplied by the size of the population. Since the mutation rate is similar for many organisms, the time needed for certain changes to arise depends mainly on population numbers.

Evolutionists also omit one key factor from their time argument - the environment. During slow evolutionary changes lasting many millions of years, environmental conditions also change. We know from the climatic history of Earth that environmental conditions have changed drastically over the last million years. Earth glaciations occurred every 100,000 years causing huge temperature fluctuations. Such drastic climatic changes must have had a dominant effect on the reproduction rate of living organisms, as opposed to any slow evolutionary changes.

Natural genetic engineering

Darwin's evolution theory has been facing more and more critics. Many biologists confronted with the incredible complexity of cells are not comfortable with the idea of changes being driven by random mutations, and propose more plausible solutions. One such solution was proposed by Jim Shapiro, from the University of Chicago, who in 1992 put forward a hypothesis that evolution was not driven by random mutations but by a process in the cell which he called "natural genetic engineering". He proposed that the cell has engineering tools such as cutting, duplicating, moving pieces of DNA and reassembling them into new genes. Because the cell can manipulate genes, evolution can progress in large steps. The most evident instruments of change are the transposons[10] that migrate from one place in the cell to another in response to stress. Shapiro claims that natural genetic engineering can produce simultaneous changes at multiple locations producing novel combinations. The latest research shows that cell's genomes are far more dynamic than originally thought and can respond very rapidly to any environmental changes.

Shapiro makes the assumption that junk DNA plays a very important role in the making of new genes and is an essential component for formatting coding information. He thinks that the genome is more like a computer containing an operating system. However he does not explain where the engineering tools come from and how they could be employed. So, effectively he proposes an intelligent design without a designer. It is true that the cell has many sophisticated tools to manipulate DNA, and genetic engineering performed in laboratories uses similar tools. But these cell tools and processes had to be designed because they could not have evolved via random mutations. So Shapiro's arguments and data support evolution directed by intelligent beings, although his proposal does not extend to this conclusion.

Comments

Darwinian evolution, despite the work of many thousands of scientists, still remains in the domain of hypotheses. Whilst enormous progress has been made in the understanding of structures and the functioning of cells and their molecular components, very little explanation has been provided on how these structures, molecules and functions have developed as the result of evolution. One cannot avoid the impression that scientists, whilst paying lip service to evolution, shy away from detailed explanations of evolutionary mechanisms. As a result of this situation, evolutionists do not feel any necessity to prove Darwin's evolution any more. It is simply taken for granted that it is correct.

It is a sad fact that evolutionists present to the public a false understanding of evolution. The evolutionary processes they describe are purposeful, intelligent, and intuitive but have nothing in common with the random mechanism which is supposed to be the driving factor.

The workings of evolution are very confusing because the results of evolutionary action, such as the resistance of bacteria and parasites to medication, have been observed. It was shown that evolution is able to utilize two or three beneficial mutations when population numbers are high. However mutations are not able to generate larger beneficial changes in organisms or make new proteins.

It is well proven that the mutations of genes take place but they are very harmful and cause many serious human diseases. To protect organisms against the effects of mutations the DNA copying process has very sophisticated built-in repair mechanisms. This proves that mutations are "not wanted" by evolution.

Sexual reproduction provides a very efficient mechanism of genetic variation. This variation is responsible for the selection of beneficial traits and enables organisms to adjust to the changing environment. Evolutionists present this mechanism as a driving force of evolution and all proofs of evolution are based on the operation of genetic variation. However, genetic variations as mutations can implement only very small changes to organisms and cannot be responsible for the development of new body plans.

Another confusing issue is the operation of natural selection. Organisms which are subject to some mutations or are changed due to genetic variability are tested by the environment and unsuitable versions are removed by natural selection. Therefore natural selection takes place, but on a much more limited scale. It cannot be responsible for making new molecular complexes or new body plans.

The evolution theory finds itself on the defensive because many scientists find it difficult to accept in the light of new discoveries. Evolution is based more on belief than on scientific proofs and is waiting to be replaced by a more plausible paradigm.

CHAPTER 11

Intelligent design and directed evolution

Paraphrasing Nietzsche we could say that "the God of evolution is dead" and now it is time to bury him and find the true paradigm of life. Before we look for new solutions let us recap what really happened on Earth.

The journey of life began about 4 billion years ago with simple, bacteria like organisms and has ended with man and his incredible brain. It started with single cell organisms and finished with 37 trillion cells in the human body.[1] The first bacteria had a genome consisting of about 5 million base pairs while man has about 3.2 billion base pairs.[2]

In the beginning there was only one species, now it is estimated that there are about 9 million different species. This means that there are 9 million different body designs, 80 percent of them still undiscovered.

So we've witnessed incredible progress in the development of life during these last few billion years, which has moved through several phases. The first phase was the arising of life represented by bacteria-like cells. Although these cells were relatively simple they already had very sophisticated energy generation and transformation systems as described in chapter 5. The second phase was the development of eukaryotic cells which have very complex genetic apparatus. These cells paved the way for the third phase, the development of multicellular organisms. Based on the body plans which originated in the Cambrian explosion, the development of life moved to the fourth phase when smaller modifications of organisms took place which resulted in new families and species of plants and animals. I believe that purposeful design was responsible for the first three phases, and the fourth phase was the result of a special process.

Intelligent design

Many scientists found that the theory of evolution could not be proven and needed to find another, more plausible explanation for the arising and development of life on Earth. As a result, about thirty years ago, the scientific theory of intelligent design was proposed which postulated that organisms could not have arisen as a result of evolution due to the complexity of their biological design. They must have been created by an intelligent designer. The intelligent design hypothesis does not identify the designer and neither does it reject God.

Intelligent design theory is gaining some foothold in our minds but very often it is identified with creationism and God, therefore its rejection is based more on prejudice than science. The first scientists who championed intelligent design focused on the anatomy of animals. This approach was correct but left certain shadows of doubt which evolutionist exploited for their benefit. A different situation exists in molecular biology, where the arising of huge molecular complexes as described in chapters 3, 4 and 5, via mutations and natural selection was just not possible. Evolutionists avoid discussing molecular biology as they have no plausible explanations of how these structures could have arisen. However, the fallacy of the evolutionary hypothesis can only be proven by the study of molecular biology. This had been done by Michael Behe, who in his book Darwin's Black Box[3] published in 1996, discussed several molecular structures and biochemical processes showing that they could not have been products of evolution. He introduced the concept of 'irreducibly complex design' where structures cannot be built using small steps. He stated that, "the story of the impact of biochemistry on evolutionary theory rests solely in the details....the biochemistry offers a Lilliputian challenge to Darwin. Anatomy is, quite simply, irrelevant to the question of whether evolution could take place on the molecular level. So is the fossil record."[4]

When we look at the last 4 billion years of life on Earth and the immense complexity of molecular components of cells, the necessity for the involvement of intelligent designers becomes evident as there is no other way to achieve such immense purposeful changes. We do not know how this design was implemented as our understanding of the workings of cells is still very limited. This involvement was not a 'one off' as described in religious texts but took place during each critical phase of the development. Michael Behe calls this involvement the 'fine tuning' of nature. He believes that this involvement was very extensive: "From what has been learned in the past few decades about the complexity of the genetic basis of animal development, it seems reasonable to think that purposeful design extends into biology at least to the level of the major classes of vertebrates, perhaps further."[5]

We can assume that the intelligent design theory has been proven beyond any doubt and evolutionists are not able to find any convincing scientific arguments against it.

The question is whether intelligent designers were involved in the design of minor classes of organisms such as families and species? I believe that another mechanism was responsible for these developments which I call directed evolution.

DIRECTED EVOLUTION

Whilst we can agree that the involvement of intelligent designers was necessary for the development of major body plans, it is more contentious to assume that they were, and still are, involved in the development of new life forms. I believe that it would not be practical for intelligent designers to be involved with the body plans of lower classes of animals such as families or species. Because of Earth's rich and changing environment, organisms must have a mechanism which enables them to adjust to new conditions and to generate new life forms to fill environmental niches. I believe that to modify existing body plans a new mechanism which forms an inherent part of the cell must be in operation, and this process is directed evolution.

The primary function of directed evolution is to provide some adjustments to existing organisms to increase their chances of survival under changing environmental conditions. The secondary objective is to generate changes to existing body plans resulting in further development of life to fill environmental niches. For example, species in the family Felidae, which include cats, lions, tigers and cheetahs, are very similar, and therefore would not require substantial genetic modifications. They could be a product of directed evolution.

Directed evolution is a process operating within the cell which was introduced by the designers. It is a built-in mechanism which responds to varying environmental conditions. To implement directed evolution the designer would have had to introduce several functions which would enable organisms to not only adapt to a new environment but also to progress further on the developmental path by generating new species. These would include genetic functions and epigenetic functions.

Genetic functions

Horizontal transfer of genes

We observe that organisms have exceptional abilities to adapt to new conditions. Simple cells such as bacteria have limited capacity to change because their genome is rather small. However they have a special built-in function which enables them to adapt because it provides an extraordinary flexibility of their genetic makeup. It is known that bacteria not only inherit genes from their mothers but accept genes from their neighbors as well and absorb genetic material from the outside. Somehow bacteria can acquire DNA from other bacteria which might have more advantageous characteristics and use it for their own benefit. This process is called the lateral or horizontal transfer of genes and played a very important role in helping bacteria to survive the most challenging conditions on Earth and to fill many environmental niches during the last 4 billion years.

Transposons

Eukaryotic cells do not use the lateral transfer of genes as this would not be an easy process for multi cellular or land based organisms. They therefore need a different mechanism of adaptability. One of these mechanisms uses special features of the eukaryotic DNA which enable significant changes of the genome without the need for mutation. These changes are possible because eukaryotic cells have larger and more intricate genomes than bacteria including significant amounts of non coding material very often called 'junk' DNA. A large part of this 'junk' DNA consists of transposons (see chapter 4).

Transposons play a very active role in the genome. They can cause rapid duplications and rearrangements of genetic information. They can add large new tracts of DNA to the genome, initiate alternative splicing of DNA segments and provide a wide range of different gene modification tools. A new genetic material can be formed from transposons as well as creating novel regulatory circuits.

Some scientists believe that transposons play a very important role in the development of life. They provide a foundation and can facilitate the changes in organisms. Some scientists call 'junk' DNA 'pseudogenes' and postulate that they could be a source of new genes.[6] The 'junk' DNA is a huge source of spare nucleotides which could be used to assemble hundreds of new genes without the need for mutations.

Other genetic functions

In the DNA of eukaryotic genes, protein coding sequences are not continuous as in the genes of bacteria, but are separated by non-coding sequences called introns (see chapter 4). Each gene may have several introns in its DNA sequence. This arrangement offers the exceptional facility to change the gene coding by removing or adding different sections of DNA sequences. Because of this facility one gene can code several different proteins. At this moment we do not know what controls this process

As was discussed in chapter 6, Hox genes provide the overall control of body plans. These genes are conserved, meaning they are not affected by mutations. However Hox genes also control other genes which describe more detailed structures of organisms. These subservient genes, when modified, could cause small alterations to body plans. As such this process could also be responsible for directed evolution.

As has been shown, animals and plants have built-in facilities helping them to substantially change their genome, or to change the expression of important genes. We do not know the mechanism of all these changes but we know that stress may play a very important role.

Epigenetic functions

Epigenetic functions cause changes to the bodies and workings of organisms without altering their genetic code. As was discussed in chapter 6 not all genes in the cell are active and some can be switched off. This is achieved by modification of the microstructure of DNA itself or the associated proteins. This mechanism enables differentiated cells in a multicellular organism to express only the genes that are necessary for their own activity. Epigenetic changes are preserved when cells divide.

There are different types of epigenetic marks and each one tells the proteins in the cell to switch off those parts of the DNA. For example, DNA can be tagged with tiny molecules called methyl* groups. Other tags can be added to proteins called histones* that are closely associated with DNA. The whole process is reversible by removing the epigenetic marks and switching the genes back on.

The epigenetic mechanism enables the cell to react to external conditions and change its performance. However, sometimes these reactions could be very detrimental causing diseases such as cancer and cardiovascular problems. Epigenetic change is a regular and natural occurrence but can also be influenced by several factors including lifestyle and environmental stresses.

Stress functions

It is known that organisms are not able to produce progeny up to their maximum capacity as reproduction efficiency is limited by environmental factors. These factors include the availability of food, water, living space, and general and climatic conditions. We should also add exposure to toxins, predators, parasites and diseases caused by viruses and bacteria. Each of these elements puts some pressure on the organism resulting in stress. These stresses could be chronic resulting in permanent damage or even death.

Cellular stress responses are expressed primarily through making stress proteins. The main purpose of these stress proteins is to secure the survival of cells. The most common are heat stress proteins generated as a result of temperatures which are outside of the normal range. As a result of stress, cells produce enzymes which perform various important functions. For example, such enzymes can pinpoint a specific part of the DNA and cut it to provide the necessary materials to make another gene. It is well proven that stress can cause epigenetic gene switching. It is well known that extended stress has far reaching consequences for the organism and could be the main mechanism which drives directed evolution.

For many years I have been interested in the effects of stress on organisms and in March 1990 I submitted a paper to the University of Chicago entitled "The role of stress mutation in evolution" for publication in *Evolution Theory*. The paper was rejected because I was not able to prove the proposed new mechanism of evolution. In my paper I put forward the hypothesis that evolution cannot be driven by random mutations but by a process in the cell itself. This process could be triggered by stress in the cell caused by external or internal factors.

The role of viruses

Viruses do not have a good reputation as they tend to be associated with diseases such as AIDS, microcephaly or cancers. Our failure to cure these diseases adds to our fear of these organisms. However, most viruses do not cause diseases and only then if a well-established equilibrium becomes imbalanced.

Viruses make their living by breaking into cells and using the machinery and energy in the cell to reproduce. Some viruses upon entering a cell make many copies of themselves which in turn infect further cells. Other viruses do not destroy the host cell but inject copies of their genes into its DNA. When the cell divides, it copies the newly acquired viral genes along with the rest of its genome. In this way the virus is copied in every daughter cell.

When the infected cell is an egg cell, viruses are copied into every cell in the organism that grows from the egg and then into all of the organism's offspring. In this way viruses become integrated into their host's genome. Once established in the host's genome, viral genes can acquire new functions, and for example, when ancestral mammals were infected by a virus that became incorporated into their genome, it helped to connect the placenta with the uterus.

Viral genes appear to be more active in the genome than original genes. For example, it was found that viral genes in the open regions[7] of DNA are more common than in the genome overall; in other words, these regions were enriched in viral genes. What is interesting is that the viral genes in the open regions of the human genome appeared during the arrival of primates[8]. It has been recognized that that infection by viruses which became incorporated into the genome considerably contributed to the development of primates.

Organisms experience a wide range of environmental factors such as temperature and nutrient supply which pose challenges to biochemical processes. As a result of changes to these factors, proteins have to adapt to new conditions. New research provides proof of how viruses shaped the development of humans and other mammals on Earth. Its findings suggest an astonishing 30 percent of all protein adaptations since humans' divergence with chimpanzees have been driven by viruses[9]. The results have revealed that adaptations in mammals occurred three times as frequently in virus-interacting proteins compared with other proteins.

A DNA sequence derived from a virus and present within the genome of a germ cell (egg or sperm) is called an endogenous retrovirus* (ERV). ERVs belong to a type of gene called transposons, which can be packaged and moved within the genome to serve a vital role in gene expression and in regulation.

Many human genes are accompanied by tiny stretches of DNA called enhancers. When certain proteins latch onto the enhancer for a gene, they start speeding up the productions of proteins from it. Scientists found that the PRODH gene which is most active in a few brain regions such as the hippocampus was enhanced in humans by a virus which infected our ancestors. This situation appears to be unique to our own species. Chimpanzees have the PRODH gene, but they lack the virus enhancer. They produce PRODH at low levels in the brain, without the intense production in the hippocampus.

Over the last 100 million years retroviruses have repeatedly integrated into the germ cells of their hosts and thus have become incorporated into their genomes. This incorporation changed the operation of the organisms bringing new functions and enhancing old ones. Viruses provide new genetic material, can modify genes and can enhance the operation of existing genes. Their effectiveness in the modification of organisms could be several orders of magnitude higher than that of mutations, therefore they are one of the most important mechanism for driving the development of life.

Mechanism of directed evolution

The process of directed evolution needs a supply of DNA material, which could be used to make new genes, and needs instructions on how to make these genes. Both of these functions could be provided by transposons. Since they make up a large part of the genome they could be a source of coding and non-coding material that enables the making of new genes and regulatory sequences. Transposons could also change the genetic makeup by adding, removing or rearranging genetic material and are able to rearrange chromosomes. They behave as mutagens but with up to 1,000 times higher mutation rates.

New research shows that eukaryotic cells have tools which enable not only genetic modifications but the generation of new genetic material and the creation of novel regulatory circuits which could lead to the arising of new species. These complex tools did not arrive via evolution but were already present in the early ancestors of eukaryotic cells[10]. This is supported by the evidence that genomes of different organisms such as sponges, worms, plants and man all have approximately the same genetic content, so their genome size does not seem to be related to their complexity.

So we have identified the tools needed to perform directed evolution processes, but the question arises of how these tools are driven. The main factor inducing changes in organisms is the environment. When we look at fossil records we find long periods of stasis where organisms did not change at all. We can also observe new species appearing in a very short space of time. This fact forced evolutionists to invent the new theory of punctuated equilibrium.[11]

The connection between transposons acting as gene regulatory agents and the environment has been well established,[12] as well as the fact that transposons can react to environmental stresses due to microclimate changes.

Since environmental stresses produce stress proteins I propose that stress proteins act on transposons and initiate genetic changes. Therefore, not only climate change, but any other stresses caused by diseases, parasites, overcrowding and lack of food could also cause significant genetic changes resulting in new species.

It is also proven that stress causes epigenetic changes which can affect body changes. However epigenetic changes operate on the existing pool of genes therefore have less of an impact on the generation of different body plans.

I believe that the generation of new species is a semi-autonomous process taking place continuously, filling environmental niches. We do not know yet how extensive the changes introduced by directed evolution could be, but they could include even higher classes of organisms such as orders and families.

The supporting evidence for directed evolution comes from observations that, faced with new environmental conditions, populations of organisms will eventually generate new species. The fastest known changes, only taking several hundred generations, have already been observed in some insects. And new research has shown that sockeye salmon evolved into a new species in fewer than 13 generations.[13] Such rapid changes cannot be explained by standard evolutionary rates.

Laboratory investigations show that in fruit flies, *Drosophila melanogaster* recombination[14] of genes increases at developmental temperatures above and below normal culture temperatures, as well as from nutritional stress in the form of starvation. Therefore these experiments provide a direct link between stress and genetic changes.

The internal restructuring of genomes takes place as a result of these factors. This process is not completely random as it provides a unique response to specific environmental conditions. It can trigger direct genetic changes or enhance the rate of mutations. These mutations do not aim to reduce specific stresses but to induce morphological changes which are then permanently fixed in the population through natural selection.

It is important to note that this process could simultaneously occur in many individuals living in the same geographical area subjected to the same kind of stress. Therefore genetic changes caused by directed evolution do not need to be present in large populations, as is the case for random mutations, but can be effective in relatively small groups. In such situations, induced beneficial changes are more quickly assimilated in the population than changes resulting from random mutations of individual organisms.

Comments

In spite of the fact that many years have passed since the publication of Darwin's theory, evolutionists have not been able to provide any significant scientific proof that this is a viable mechanism responsible for the development of life on Earth. The weakest part of this theory is the explanation of the origins of life. The proposed hypotheses are so preposterous and are so much at variance with the laws of physics that they cannot be seriously considered.

This leaves us with the only viable solution which is the theory of intelligent design, stating that a designer was responsible for the arising and development of life on Earth. However, intelligent design does not explain the appearance of millions of species and therefore we have to look for another mechanism. This mechanism is directed evolution which is a part of intelligent design. It was introduced right at the beginning with eukaryotic cells, although some elements were already present in bacteria. This mechanism is based on the extraordinary properties of transposons which are able to extensively modify DNA sequences to generate new genes and gene regulators.

It is known that the operation of transposons can be triggered by environmental stresses and I propose that stress proteins are the link between environmental stresses and genetic modifications. This link still has to be established, but it is only a matter of time.

Viruses play a special role in the development of life. It has been shown that they are able to enter DNA and change genomes. Some viruses could generate new genes and some could be responsible for switching existing genes and in this way contribute to directed evolution.

Intelligent design has been proven beyond any doubt. The only question which remains to be answered is, "Who was the designer?" I will try to answer this question in chapter 13.

CHAPTER 12

Engineering design

Engineering approach to biology

Until recently, when our knowledge of the construction and operation of many essential molecular complexes such as ribosome, spliceosome, ATP synthase, electron transport complexes, photosystem units, etc. was very limited, one could tolerate statements such as 'the structure of cells is very simple'. However once our knowledge of these molecular complexes reached a level of fair understanding, these statements became no longer plausible. These molecular structures operate on similar principles to machines made by man with the only difference being that they are made from organic compounds and not from metals or plastics. Since we know so much about the structure of these complexes we can start applying engineering knowledge and rules to explain their functioning.

It is the lack of understanding and appreciation of engineering which is at the root of many statements implying that highly intricate constructions would be able to arise through random mutations and natural selection. Scientific communities, especially biological communities, do not realize how difficult it is in real life to make things work as they should. Modern aircraft or computers require many millions of hours of design and testing to reach the required level of performance and reliability. In industry nothing comes easy. Everything we have around us is the result of intelligent design and intelligent manufacturing. We know intuitively that a broken tea cup is not going to mend itself even if we wait a billion years, that an aircraft is not going to be built by monkeys even if they have all the necessary materials, but somehow we are induced to believe that in nature everything is possible, that evolution can perform miracles, it just needs a lot of time. Some solutions proposed by evolutionists are simply preposterous. In the engineering world they would not last a second but in the biological milieu they are considered very seriously and eventually become part of the established 'truth'. This situation is allowed to happen because molecular biology being so complex is not only difficult to comprehend by the general public but its implications are ignored by many evolutionists.

In this chapter I will look into biological structures and systems from an engineering point of view and I will try to describe what principles were guiding the design and development of these biological assemblies. There is no reason to treat biological systems and subassemblies differently from mechanical or electronic designs, and no reason not to use the same rules which apply to engineering. When we look into the whole process of the development of molecular components, from the most basic to the most complex cells, we notice that it follows sound engineering rules. These rules include unity of design and manufacturing tolerances.

The rule of engineering – unity of design

The first rule of engineering is 'not to re-invent the wheel.' So whenever possible, any new design should use parts and subassemblies from existing designs. Even when the smallest, most insignificant component has been working well for many years and is proven to be reliable, there is no reason to replace it with a new design. This is very important, especially in the aerospace industry where the pedigree of the design guarantees the reliability and safety of the aircraft's operation. Therefore, an engineer would always try to use or modify an existing part when creating a new device or machine. Only when an old part cannot meet new requirements a completely new part is designed. This approach reduces risks linked with the introduction of new unproven parts, reduces testing, verifying new designs and keeps manufacturing of these parts simple. It is common knowledge that a new product carries much higher risks than a proven product and therefore has to be extensively tested. Engineering differs from science in its approach because it does not introduce new things for the sake of novelty and being first. Engineering must provide things which work well and reliably. So in the real world, progress and changes originate from new demands for better performance.

A similar approach can be seen in the development of biological systems. First it is noticeable that all new solutions always include components of some previous designs. This is more visible on the molecular level. Therefore, one can build a tree of progress where new designs originate from earlier designs. This tree would be very similar to the evolutionary tree and in a sense could support a form of evolution, but not Darwin's evolution.

First, let us look at the key component of life, DNA which is built from 4 nucleotides which are all right handed isomers*. What does this mean? Normally in nature, molecules having exactly the same chemical composition and properties may exist in two forms, one molecule's structure being the exact mirror image of another molecule, just like the left hand is the mirror image of the right hand. If DNA was formed via the evolutionary path, any natural random process should generate an equal number of left and right handed DNA molecules, but this is not the case. All the DNA structures and coding systems are the same in all organisms which have existed over the last 3.5 billion years. All of them are right handed. Therefore, it was an engineering decision to use one nucleotide orientation for all life designs. This means that life began not only once but also as a single cell event, because any multiple origins would have a mixture of left and right handed molecules.

All organisms on Earth are built from the same twenty amino acids. Furthermore, all these twenty amino acids belong to the left hand class of the same optical isomer. The fact that all amino acid molecules used in living bodies are left handed cannot be explained by evolutionists and it is wisely omitted from all books on evolution. Since both classes of amino acids could produce the same proteins there is no reason why in another place and time random selection did not pick up right handed molecules.

As mentioned, an engineer would use existing proven working parts for use in different functions. For example, nucleotides which are used for DNA coding are also utilized as common blocks for building different RNA structures as described in chapter 3. While the messenger RNA performs coding functions, the ribosomal RNA is the main building block of a ribosome and therefore performs a structural function. The spliceosome, a molecule critical for transcription, is also made of nucleotides. Besides being used for coding and building large complexes, nucleotides are also used to build ATP – the universal energy currency molecule. So we can see that nucleotides are the most frequently used building blocks essential for life. What is important is that they are used in different structures performing completely different functions. Only an engineering mind would offer such solutions.

When we look at the structure of hemoglobin, a critical molecule in our blood which carries oxygen, it includes the heme group which in the middle has one atom of iron, Fe. However when we replace iron with magnesium, Mg, we have a chlorophyll molecule (see Fig. 5-2). So both metal binding molecular structures are identical and the only difference between these two molecules performing completely different functions is a different atom of metal. The selection of the heme structure to perform two different functions could only be an engineering decision.

To summarize we can conclude that only an engineer who understands the advantages of the existing design would use this knowledge in a new design. Only an engineer would introduce the commonality of basic construction subunits and would not try to invent new components. Only an engineer would use the same complex molecules such as the heme group to perform completely different functions. Only an engineer would use one common cell in different organs of animals performing different functions. In reality, wherever we look, we see this engineering approach to the development of all life on Earth.

A special engineering solution

As was discussed in chapter 4, evolutionists had to admit that the arising of the eukaryotic cell happened only once. Therefore, they proposed that the eukaryotic cell was a 'chimera' – a child of two completely genetically different parents. In Greek mythology, the chimera is a fictional beast composed of parts taken from different animals such as a lion, a goat, a snake, etc. So what evolutionists propose is that the eukaryotic cell came about as the result of two different bacteria joining together resulting in a completely different daughter cell.

The chimera solution makes sense from an engineering point of view. When we have two or more already proven simple devices and we want to build a more complex piece of equipment providing substantially improved functions of the simpler devices, the obvious solution always used by engineers would be to join the simple devices together. However, the joining process is not as simple as assumed by evolutionists. A proper interface between the original devices has to be designed, additional parts have to be added to facilitate a smooth transition and a new control system must be designed as the new device is more complex and the interactions within it are more intricate. This is precisely what is observed in the eukaryotic cell which has many additional unique structures such as a nucleus, cytoskeleton, Golgi apparatus and other organelles which could not have been taken from one of the progenitors since no bacteria had them before.

So we could argue that evolutionists who, even reluctantly, accept the chimera solution move away from the idea of small step changes and in reality accept engineering design, but call it symbioses.

The rules of engineering - tolerances

If we could closely inspect several machines manufactured in the same way and performing the same function, we would find that they would not be identical. This is due to imperfections in the materials and limited accuracy in the manufacturing processes. So the manufactured parts never have the exact dimensions as described in the design drawings. To ensure that different parts fit and work together, an acceptable range of dimensions must be defined. Therefore, all dimensions on design drawings are given with specified tolerances. The designer determines these tolerances from his manufacturing knowledge and from his understanding of the interaction between two or more different parts working together. The designer ensures that the parts will perform their functions if their manufactured dimensions are within the specified limits. The implication of the existence of tolerances is very significant. It means that in engineering there is no possibility of changing a design using gradual steps.

This can be explained using the famous example of the evolution of the watch. Let's assume that we want to change a small part of a watch, for example a gearing transmission by increasing the number of teeth on one cogwheel. To engage properly the teeth on both wheels must be identical and therefore the radius of the changed wheel must be increased to accommodate the additional tooth. However as soon as the radius increases, the gear will stop working because the wheels will lock. So to increase the radius we must simultaneously increase the distance between the wheels, so we actually have two dimensions to change. Even if the radius and distance of the wheel can be changed, a gap will appear where the additional tooth should be added. So again, the gear will fail to work until this additional tooth appears with the correct size to be able to engage.

This illustration shows that it is not possible to change a simple thing like a gear ratio in small increments as any small change of even one dimension outside tolerances will cause the gear to fail. So the process of changing a gear ratio would not meet the principle of evolution which states that at each stage of development an organism must be fully functional.

In complex mechanisms it is impossible to change a component in small steps because of its interaction with other parts. In practice the whole device would have to be redesigned from the beginning, meaning that several dimensions would have to be changed simultaneously. The same constrains would apply to molecular systems which operate like electro-mechanical devices.

Tolerances in digital systems

While mechanical devices can have their dimensions changed within their tolerance limits and still function correctly, different conditions apply to digital systems. A digital system has a digital code which does not tolerate even the slightest changes. For example, a simple code could look like a series of "zeros" and "ones" e.g. 1100101111001000. Even the smallest change in the code, like replacing one "0" with a "1", would result in the faulty operation of the whole system. You could check this easily, for example by typing a URL address in your internet browser. Try to change one letter or put a "dot" in the wrong place and you will never get to the intended website. So a digital system which offers many advantages and benefits is not error tolerant. To avoid malfunctioning, digital system signal levels must be much higher than analogue ones. They must have built in redundancies which can code the same information using different coding sequences, or have special fault detecting systems which weed out signal errors.

Tolerances in molecular machines

A similar problem exists in biological machines. As we know in DNA information transfer is based on digital coding. It is not the same coding as used in computers, but nevertheless the principles are the same. DNA provides instructions for building proteins using three letters codes, each corresponding to a different nucleotide. The smallest possible change of the coding is the replacement of one letter by another letter which would result in a different amino acid being used in a protein chain. What are the consequences of using a different, randomly chosen amino acid? As we discussed in chapter 3, a protein chain is folded in a very convoluted way forming globules, cylinders or many other shapes. Only a specific, small part of the protein interacts with other proteins and enzymes, and this part must be properly exposed by being outside of the folded structure, because the intermolecular and atomic forces only operate over very short distances.[1] Therefore, when a different amino acid is placed in the chain this could drastically affect the folding of a protein, resulting in a completely different shape. In such a protein the interactive area could be located in a different position, possibly inside the fold and in this case such a protein would not function. It is possible that in the mutated protein other interactive areas could be exposed, but this would also require changes in the other interacting proteins which would also have to be redesigned. So mutations must affect at least two different proteins in precisely the right way and at the same time and as such, these mutations cannot be random. In practice such a change would require both proteins to be completely redesigned.

However the good news is that biological components also have defined tolerances. A protein can perform its function even when some amino acids have been changed. These changes are called neutral mutations, and as we know about 30 percent of all mutations are neutral. Therefore we could say that a given protein structure has tolerances within which it can operate correctly in spite of some structural changes.

It is interesting to notice that man made components do not have such capabilities. To protect a system against malfunction, engineers introduce spare components which, in case of an error, replace faulty components.

Cell control system

The most important feature used in engineering is control. Without control it would be very difficult to perform any function or operation. The purpose of control is to check that the performed function progresses as it should. Any function or process is normally subject to interfering elements which would affect the outcome of the operation. It could be unpredictable changing environmental conditions, variable supply of materials or noise in the system which limits the accuracy of sensors. So the control system is needed to remove unwanted and unpredictable variables and influences to stabilize the operation. For example, during manufacturing processes we have to know how many objects should be made and when to stop the process. Such processes can be regulated by the most common type of control which is based on negative feedback.

Principles of negative feedback

In our daily lives we employ a vast amount of control. Sometimes we might not realize it because our body is doing it automatically without our awareness. For example, let's assume that we are driving a car and we want to turn left. First we start turning the steering wheel in an anticlockwise direction and the car starts changing its direction. At the same time we observe the position of the car and when we find that the car is moving too much to the left we slightly turn the steering wheel in a clockwise direction. We check the new direction and might implement further corrections by turning the steering wheel. We do this automatically because we have learned how to drive a car. However if we are driving for the first time we would notice that the car's movement widely oscillates too much to the left and too much to the right and we would have to keep correcting it. This type of control is called negative feedback.

Negative feedback is implemented everywhere: in industry, in our everyday lives and in nature. For example, a robot needs very precise information about where its arm is and with what speed it is moving. To implement this control we need several sensors telling us about the position of the arm and its movements in 3D space. Sensors send information to the control system which compares the position of the arm with where it should be and generates a correction signal. Without this control a robot would not be of much use to us.

The same type of control is extensively applied in biological systems. Any manufacturing function performed by the cell must be controlled otherwise the organism would waste a huge amount of energy and material resources. This control is based on negative feedback. So the feedback control lies at the heart of any biological operating system. To understand it better, let's consider the following illustration:

When a cell needs new proteins it generates a demand signal saying, for example, 'produce 30 proteins'. This signal is sent to the appropriate molecular units which manufacture these proteins. However without any control the cell would produce these proteins nonstop, until all available raw materials in the cell have been used up. A diagram of such a system is shown in Figure 12-1(a). Such a system is not much use to a cell.

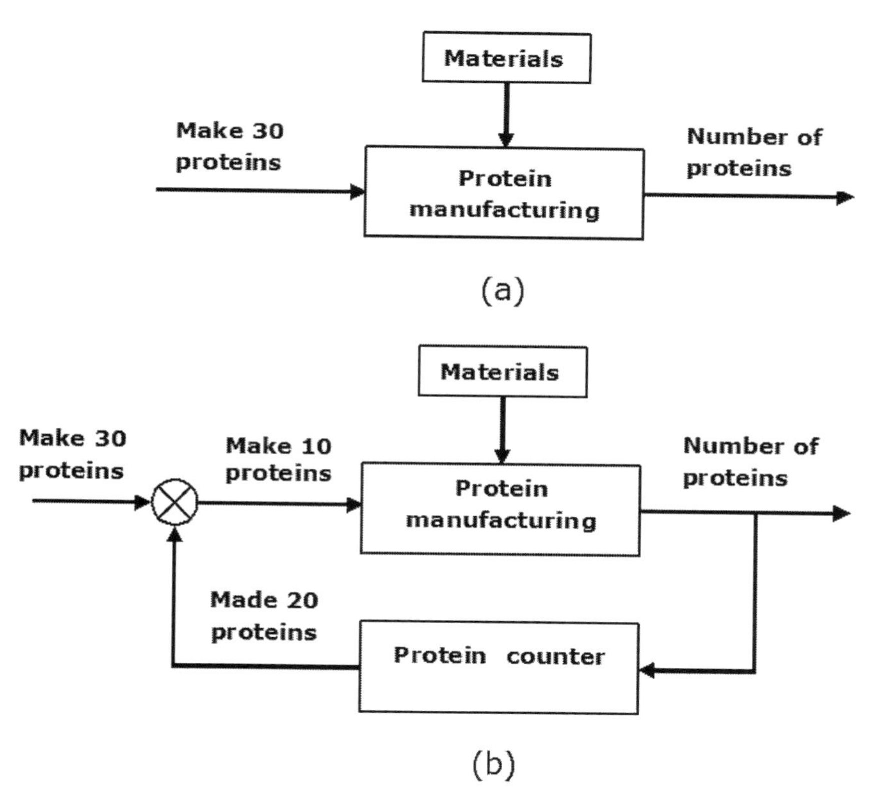

Figure 12-1. A diagram showing the principles of negative feedback control.

The production process must be modified by introducing a negative feedback control loop. To implement it a sensor is needed which is able to identify the produced protein when it comes out of the unit. Next, a counter is needed which will be able to count the number of produced proteins. And finally the signal from the counter is sent to the demand unit telling it how many proteins have been made. The diagram shown in Fig.12-1(b) illustrates that the unit has just made 20 proteins and the signal is sent to the input device which subtracts this amount from the 30 original units and provides a new demand signal saying to produce 10 proteins, and so on. When 30 proteins have been generated the demand signal is zero and the unit stops manufacturing.

Such a system must be used in every cell production unit controlling the flow of materials and demand signals. This means that the cell control system is significantly more complex than was assumed just a decade ago.

Model of the cell control system

The cell is a very complex manufacturing conglomerate with thousands of different operations and hundreds of operational units. Like every manufacturing enterprise it must secure and manage its production by using a proper control system otherwise it will not achieve stability and continuation of existence. To help us understand the complexity of the cell's control system let us consider how such a very simplified system could look. A block diagram in Figure 12-2 shows the most critical cell functions such as sensing, energy production, maintenance, reproduction and transport. Each of these functions could be split into several sub-functions but we still do not know much about the functioning of the cell. Each of these functions perform very important operations which depend on other functions, therefore there is network of interactions and interdependence.

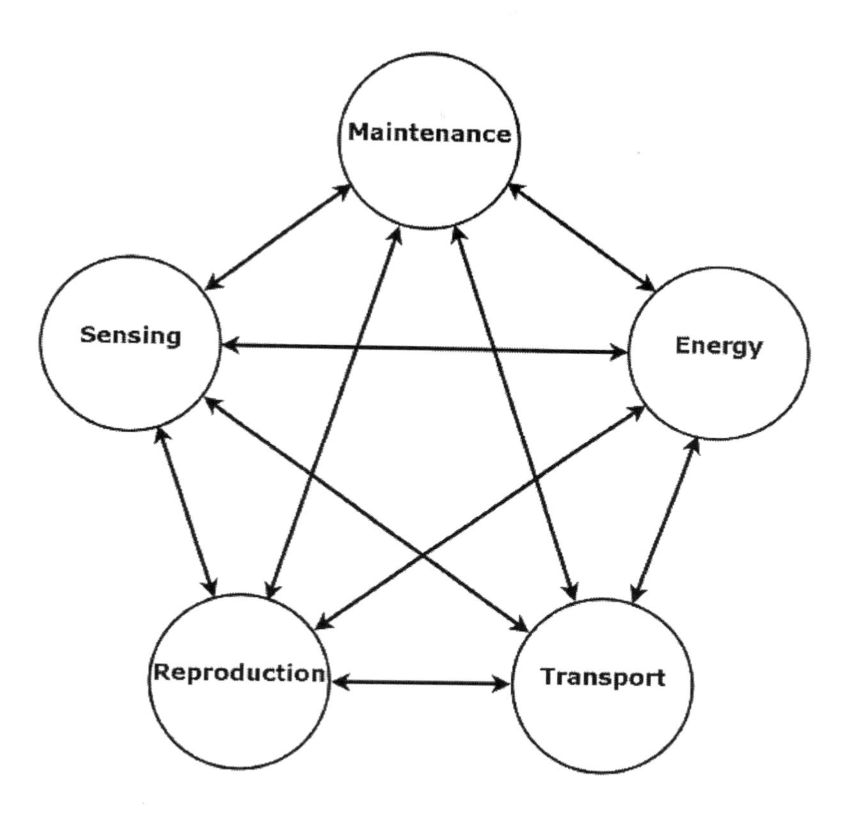

Figure 12-2. Diagram of the cell control system.

At the heart of any control systems are sensors providing feedback in each control loop. The cell needs a variety of sensors to provide information about its environment such as temperature, light, oxygen, nitrogen, carbon dioxide, water, salinity, nutrients, metals and many other elements needed by the cell as well as to give warnings about potentially toxic compounds. There are special sensors which detect other similar cells or hostile organisms. Sensors also provide information about what is happening inside the cell. Is there enough water, nutrients, metals, oxygen, carbon dioxide, etc.? Are any waste products to be removed? There are sensors telling the cell if there is a need to make more amino acids, proteins or larger molecular complexes like ribosome, RNA protease, ATPase, etc. There are sensors detecting malfunctioned molecular complexes and proteins and making decisions to break them down.

Sensors interact with all other functions of the cell. They provide information which is critical for the cell to make various decisions. If there are enough nutrients and the correct environmental conditions they provide signals to start the reproductive cycle. They interact with the reproductive process controlling manufacturing of proteins and other molecular complexes. They control the transport of water, nutrients and elements through the membrane and block transport when the cell does not need a particular compound. In the same way, if there is any waste to be removed it must be done in a controlled manner by opening and closing the correct passages in the membrane. Sensors are essential for cell maintenance - manufacturing of replacement proteins, molecular complexes, destruction of non-functioning molecules and removal of waste. In the photosynthesis process, besides sensors to detect water, carbon dioxide and oxygen there are also light detecting sensors which guide organisms to locations with good illumination. Sensors also detect the demand for energy in different parts of the cell and send signals to the energy generators to provide more ATP. Every year we learn more and more about cell sensors and the above list is far from comprehensive. Many biological sensors are so ingenious that researchers are investigating how they might be used in medicine and industry.

Energy generation via the respiration process is one of the most critical functions of the cell. The main product of this process is ATP generated by the proton pump. This function provides energy to all other processes in the cell and any interruption in its production would result in instant death of the cell. Energy must be delivered instantly, on demand, otherwise other vital functions could be disrupted. It is needed for many chemical reactions and building of new components as well as for physical movement of the cell and transport of compounds within the cell. This function must be controlled by a very sophisticated control system supervising the flow of electrons in the electron transport chain, monitoring proton potential and securing production and transport of the correct compounds. It receives signals from sensors detecting energy demand and receives feedback telling the cell that enough energy has been delivered.

The cell maintenance function is provided by a large and complex factory making all the necessary building blocks for the cell as well as building the factory itself. In cells, especially bacteria, there is very little storage space to keep spare materials, therefore a very efficient 'on demand' supply system must exist. When there is a demand signal for a certain protein or metalloproteins to be made, the factory starts building the necessary amino acids selecting the correct materials which have to be imported from the outside. Therefore, these demand signals arrange transport of the necessary materials into the cell. Since all protein building information is in the DNA a complex module, RNA polymerase, has to find where the genes with the relevant instructions are placed. Once enough mRNA copies have been made, RNA polymerase must receive a signal that no more copies are needed. Next the messenger RNA must be guided to a ribosome which must receive a control signal to start manufacturing the needed protein. The ribosome selects amino acids coded with tRNA and puts them in one long chain. When the protein is made the ribosome must receive a signal that the task is completed. The maintenance function requires many sensors and extensive interaction with other cell functions including helping to fold proteins using special chaperone proteins.

The cell maintenance function not only makes new molecules but also removes old ones. Proteins or larger molecular complexes which are damaged, not working correctly or not needed must be removed. However they are not wasted but decomposed and basic materials are reused.

Without the reproductive function the cell would die and disappear, so this function secures the continuation of life on Earth. The decision about starting the reproductive cycle is made using information from sensors telling the cell about the condition of the environment and available nutrients. When these conditions are right the cell sends a signal to DNA polymerase to start copying genes and signals to the nucleus to make new proteins, enzymes and ribosomes. At the same time the cell sends signals to the membrane to grow. It sends signals to start producing amino acids and then to copy organelles such as the nucleus, cytoskeleton and mitochondria as well as large molecular complexes like ribosomes, RNA polymerase, enzymes, ATPase etc. in the cell. The manufactured components must be placed in the correct location in the cell. When the manufacturing process is completed it sends a signal for the cell to start division. Therefore the reproductive function interacts with all other cell functions.

Transport in the cell is a very advanced operation. It consists of transport through the membrane and internal transport. Transport through the membrane employs many special proteins which control the passage of hundreds of different products. There are importing and exporting molecular complexes operating under the control of other functional modules. Many of these functions require energy which comes in the form of ATP.

Internal transport moves materials inside the cell. Amino acids must be delivered to ribosomes in different parts of the cell to make proteins. Proteins and enzymes must be delivered to the right locations at the right time. This internal transport uses special ATP energized motor proteins moving goods along the cytoskeleton like cable car lines. This transport requires control signals providing destination addresses of materials and signals demanding ATP which provides energy for transport. Cell transport is not a chaotic movement of materials but a highly controlled system comparable to railroads or motorways.

There are thousands of different processes in the cell, but we have to remember that each process needs a sensor, control signals and a feedback loop. There must always be a demand signal, material gathering signal, completion signal, selection of correct manufacturing complexes and enzymes signals, end of manufacturing signals, transport directing signals and ending of demand signals. The complexity is mind boggling. We have to remember that the sensors are made from special proteins and that the signals are carried out by proteins, this explains why more than 100,000 different proteins are used in the human body.

The system shown in Figure 12-2 looks very intricate but in reality it shows only the main areas of cell control. In practice this system is several orders of magnitude more complex and until we discover its components we will not be able to understand life. In order to simulate the operation of this system we would have to use a large computer but the cell can do it without any brain.

Our knowledge of the cell control system is very minimal. The control system in the bacteria *Caulobacter* was investigated by Prof McAdams[2] and was described in chapter 3. His observation was that even the simplest bacteria have a control system which is similar to engineer designed control systems. It is worth adding that Prof McAdams' work is just 'scratching the surface' of our understanding of the cell's control systems. One would expect that a much more sophisticated system operates in eukaryotic cells, but this is work for the future.

One could draw the conclusion that there must exist an extremely complex mechanism controlling the structure, operation and reproduction of the cell which is not coded by DNA, although the proteins used for control are coded by the genes. This makes sense because this mechanism controls the workings of DNA itself. So we have some sort of autonomous, independent control system which is copied during the cell's reproduction process and is then transmitted to the next generation of cells.

Increase in complexity

When we look around us we notice that our things are getting worn out, stop working and eventually we have to throw them away. All man-made things, with a few exceptions, disintegrate and stop being useful. This is a natural and logical way in which the universe operates. Our experience is confirmed by scientific laws, especially the second law of thermodynamics. This law simply states that energy or temperature in a closed space will equalize and any ordered system will eventually change into a disordered system.

However there is a way to increase complexity whilst at the same time obeying the law of thermodynamics. This method requires two ingredients: energy and information on how to use this energy. For example, let us consider a factory manufacturing cars. This factory is producing very complex, ordered products. But to produce them it needs information on how to make them. So, the factory is designed and all manufacturing processes are defined. Next, all the machinery and tools are acquired and workers are trained on how to use them. However, to start manufacturing, an essential ingredient is required– energy, which drives the machinery and the processes. This is the only way of increasing complexity in the material world.

The second law of thermodynamics suggests a progression from order to disorder, from complexity to simplicity in the physical universe. Yet biological evolution, which involves a hierarchical progression to increasingly complex forms of living systems, appears to be in contradiction to the second law of thermodynamics. However, evolutionists postulate that the complexity of biological systems can increase on its own by means of random steps. Evolutionists assume that biology does not have to obey the laws of physics or rather they use the explanation that biological systems are driven by energy from the Sun, therefore the system is open and the laws of thermodynamics do not apply. This is an erroneous explanation because energy on its own is not able to increase complexity and order. Energy has to be directed to appropriate channels through instructions and special mechanisms or devices, therefore information is necessary prior to the use of energy.

Living organisms, on the other hand, comply with the second law of thermodynamics because they use the Sun's energy for growth and reproduction and this energy is directed and transformed using instructions in the cell. So the cell is the depository of the necessary information. However this information was generated during the design stage of the cell.

So the increase of complexity is driven by the information. For example, if in the above mentioned car factory the management is not making any decisions, the factory will operate less efficiently, producing more waste and scrap. Even such a simple man made structure cannot operate without intelligent input.

Body plans

Are biological designs similar to engineering designs? Do 'blueprints' of animals' body plans exist? These are very frequently asked questions that I will try to answer. We tend to assume that any machine, including biological machines, should have a detailed definition of its design. To illustrate this point let us look at how man made machines are built. For example, a car is made on an assembly line where individual parts are added together until the car is ready to drive off the line. The parts could be manufactured in several different places such as different countries, but this does not matter as they are made according to very detailed drawings and manufacturing instructions. Once they are brought to the assembly line they are put together according to the detailed assembly drawings and instructions. The manufacturing procedure is very simple although the parts could be very complex. This type of manufacturing is based on the high accuracy of the design and on making parts within the pre-set tolerances so they will fit together regardless of where they have come from.

However, biological organisms are not made that way. Multicellular organisms are not assembled together from individual parts but grow from a single mother cell. This cell has a full set of instructions on how to make the final body. So the biological growing process is very different from the manufacturing engineering process.

Growth is based on the multiplication of cells which divide in a controlled fashion. The necessary information to grow an organism is similar to the information needed to make cars, viz. material and dimensions of parts and position of these parts must be known. However the assembly of organisms proceeds in reverse order to the assembly of cars. During the early stages of the embryo's development precursors of all major organs are formed which are placed in the correct position and order. This could be compared to a toy car on the assembly line. So this toy car already has every part in the right place, although the parts could be underdeveloped. To follow this illustration, the toy car starts growing, developing a bigger, more complex engine, gear box, wheels and other parts with electrical wiring already in place. The toy car changes slowly into a real car and at this stage drives off the assembly line.

This illustration is very simple but embryo development is a mind boggling process. The problem is that information about materials, dimensions and position has to be applied simultaneously during the growing stages. The whole process is controlled by the Hox genes already mentioned in chapter 6. These genes provide the overall layout of the body, defining the position of the head, main organs, limbs etc. Each of these genes switches on other genes which provide more detailed information about the position and materials of different tissues needed to make these parts. In the human body there are about 250 different types of cells such as muscle, neurons, skin, liver, etc. Therefore each stem[3] cell must have instructions on how to change into the right tissue cell and must know its position with reference to other cells. At this point our knowledge is a bit patchy.

The embryo applies several different methods to provide these instructions. For example, it uses morphogens which are chemical compounds excreted by the reference cells. These molecules of morphogens diffuse in the embryo space and their concentration varies with distance from the reference cell. Stem cells detect the concentration of morphogens and change their structure to another pre-determined type of cell. Other methods use direct interaction between cells during which chemicals are transmitted directly from cell to cell providing the relevant information.

Man made parts are relatively simple, made from one or a few materials, and the complexity is obtained at the final assembly. By comparison human organs are extremely complex right from the beginning, where for example the human eye assembly is controlled by about 2,000 genes. Because there is no separate assembly stage, embryo cells whilst multiplying must be placed not only in the right place but must cooperate with neighboring cells and organs. For example, the human blood circulation system is not made in parts bolted together, but as a whole unit and must span the whole body and work with all organs. The same applies to the nervous system and lymphatic system. Therefore there is a built-in control system which determines the interactions between different cells of the same organ. In the human embryo approximately 3,000 gene regulatory compounds are at work, and their operation is still the biggest mystery.

So the question still remains, where is the 'blueprint'? The animal body design does not resemble engineering design because it is based on different principles. It is a dynamic system based on thousands of assembly instructions making sure that all the cells are of the right type, are placed in right position and correctly interact with other cells.

In order for these instructions to have been prepared there must have existed an overall design. But this design was in the domain of the designer and was not passed to the cell. What is provided in the cells are only the assembly instructions. A similar situation exists in the engineering domain where drawings are prepared by the designers but are then passed to manufacturing engineers who prepare detailed manufacturing instructions. So for example, on the car assembly line, operators follow written instructions but they do not need to see the design drawings. In conclusion, design blueprints do not exist in the cell, only the assembly instructions.

Comments

The principles of design of biological parts and bodies do not differ from engineering design. As in engineering, the same biological components are used to build completely different parts which perform different functions. Biological components must be built with certain tolerances, otherwise the proteins will not work together.

The most important part of any engineering system is control which secures its correct operation under varying conditions. This type of control is based on negative feedback. Very similar controls operate in biological systems and they encompass all cell functions. Without such controls the functioning of organisms would be impossible.

The increase in complexity of organisms by random processes is against the laws of physics which are misunderstood by evolutionists. Energy is needed to increase the organization of a system, and information directs this process.

The building of organisms is driven by assembly instructions which are in each cell. These instructions are similar to manufacturing assembly directives. However instructions in the cell do not include blue prints for the construction of the organism. These blueprints remain with the designer.

CHAPTER 13

The greatest experiment

In previous chapters I presented ample evidence that evolution has not been able to provide any convincing scientific arguments supporting its story. It appears that recent biological discoveries make such explanations even less plausible.

A more credible solution is the Intelligent Design theory which was proposed about thirty years ago. Its main canon is that life was originated by an intelligent designer but it does not identify the designer. However, it is important to many people to know who the designer might be. Intelligent Design is rejected by academia, not as a result of scientific arguments, but because it is assumed that the designer must be God, therefore it is no different from other religious explanations.

In this book I have discussed the events which took place in our part of the Universe which led to the arising of life and man on Earth. These do not look like chance events, but have the appearance of a well planned experiment. I search for scientific explanations based on the available scientific data. This chapter will review the whole sequence of events described in this book, ending with the arising of man and identifying who the designer could be.

The sequence of major events

The arising of intelligent life on Earth took about 4 billion years and was the result of several well prepared activities. Each of these steps was critical for the development of intelligent life, being part of a long chain of events.

Life on Earth was the result of the implementation of these steps in the correct sequence and at the right time.

In my opinion these are the major events critical for the development of intelligent life on Earth:

- Selection of a suitable Sun in the galaxy.
- Selection of a suitable planet having an appropriate size, gravity, tectonic plates and magnetic field.
- Modification of the Earth's orbit to secure an average temperature in the range 10 to 20°C.
- Delivery of the right quantity of water (too much would result in no land).
- Arising of life with a complex energy generation and utilization system.
- Arising of the eukaryotic cell, the building block of multicellular organisms.
- Arising of multicellular organisms during the Cambrian explosion.
- Arising of *Homo sapiens.*
- Development of the human brain.

The above listed events took place only once in the history of Earth and they were absolutely necessary for intelligent life to arise on Earth. If just one event was omitted, intelligent life could not exist. Calculating the probability of all these events happening by chance is impossible, but we can estimate that it is practically zero.

Even a perfunctory examination would show that these events would have to be well coordinated because the timing of these events was critical for the success of the whole enterprise. They could only have been performed as a planned experiment in generating new life in the Universe, so they must have been planned and executed by intelligent beings.

Special Earth

The cosmos in general is not such a good place for life to prosper. We can now see for ourselves, when we arrived in the era of space travelling, that even a comparatively short journey such as sending man to the Moon was a very challenging task and has not been repeated since 1972. Sending man to Mars will be an even more demanding task and it is uncertain when this will become feasible.

The cosmos is very hostile to our type of biological life because of two important restrictions: a dangerous level of high energy radiation on the one side and low temperatures on the other. In the centre of our galaxy there are many stars emitting gamma rays, X rays and high energy particles which could destroy any living organisms. In practice, there is no protection against them except heavy lead shields. The only viable solution was to find a suitable planet far from the sources of radiation.

But even our Sun emits particles which could destroy life, so Earth requires a special magnetic shield to deflect these particles. There is also electromagnetic radiation such as UV radiation which can damage DNA molecules. So, as we know, Earth needs an ozone layer to reduce this radiation.

Earth is like a nuclear shelter equipped with multiple shields providing protection for biological life. It is most likely that there is no other planet in our galaxy which could fulfill all these requirements and Earth was destined to become the first "test tube" in this cosmic experiment. Hence Earth, right from the beginning, was prepared to harbor life.

Temperature is an equally essential factor for life. Water based biological organisms can survive only in a very narrow range of temperatures. We exist at the bottom of the Universe's temperature range, between 273°K[1] and 322°K, while temperatures in the Universe extend from 0°K up to several million degrees. Placing Earth on the correct orbit and maintaining the right temperature was essential to maintain life. To do this, multiple temperature control mechanisms were introduced including tectonic plates and carbon dioxide circulation.

The existence of water is another critical requirement for life. We are talking about just the right amount of water. Too little water would evaporate and eventually escape into space, and too much could result in deep oceans covering almost 100 percent of the Earth's surface. While this would make life possible, the development of advanced animals and humans depends very much on the presence of land masses. Water was not available just 'around the corner' but had to be brought from far away. Aiming a water tanker with at least 1.3 billion cubic kilometers of water to hit Earth was truly an engineering feat!

The genesis and development of life

It was discussed in chapter 3 that even the simplest biological organisms such as LUCA[2] were from the beginning very complex biochemical factories employing several thousand proteins involved in thousands of chemical reactions. Processes such as photosynthesis and respiration, which existed right at the beginning of life on Earth, employ, even by present-day scientific knowledge, mind boggling complexities. For example, the discussed mechanism of splitting water into oxygen and hydrogen is the result of an ingenious invention which, even now, escapes the grasp of scientists.

Since it would be impossible for such complex life to evolve from inorganic matter the only plausible solution is that it was implanted on Earth. Life was designed and cells were made by extraterrestrial engineers and delivered when Earth became habitable. The cells could have been delivered in special containers which were sent to our planet. The containers landed in oceans where the content was released.

The first organisms which arrived on Earth were cells similar to present day bacteria and included cyanobacteria which were responsible for the generation of oxygen. Further development of these organisms was driven by several mechanisms. The most important were bacteriophages or viruses which replicated in bacteria and could modify bacterial DNA (see chapter 11). Improved genes in some bacteria could be transferred to other bacteria using horizontal gene transfer mechanism.

This mechanism makes the direct transfer of genes from one individual to another possible, as well as picking up DNA material from outside of the cell. These mechanisms enable bacteria to adjust very quickly to different environmental conditions and fill any niches which become available. Now we find bacteria living in all conditions on Earth. From alkaline to acid lakes, from the Arctic to underwater volcanic vents. Bacteria are able to colonize all living organisms where they play an important part in their metabolism. We do not know how many species of bacteria exist, but some estimates are in the range of a billion.

After about 2 billion years, when oxygen in the atmosphere increased to sufficiently high levels, Earth was ready to accept more advanced life. Suddenly the eukaryotic cell appeared; a new building block that eventually enabled the construction of multicellular organisms. It is most likely that progenitors of the eukaryotic cell were also delivered to Earth in containers.

Eukaryotic cells have far more complex genetic modification mechanisms than bacteria. These mechanisms, which include transposons and introns, allowed organisms to modify their genes and adjust to different environments. Eukaryotic cells were also subjected to attacks by viruses which changed their genes by injecting new DNA materials.

For the next 1.5 billion years life in the oceans was dominated by single eukaryotic cell organisms belonging to the kingdom Protista. Some protists such as algae perform photosynthesis, others are predatory eating other protists and bacteria and some scavenge for dead organisms. It is estimated that at present about 250,000 species[3] belong to this kingdom. It is interesting to note that although eukaryotic cells had all the capabilities of becoming multicellular organisms, they waited so long for the next development phase. This delay was caused by Earth's climatic instabilities. Earth, between 750 and 550 million years ago, passed through several glaciation stages, some being so severe they caused all the oceans to freeze. Only when the Earth's temperature returned to normal could further development of life proceed.

This next phase was the Cambrian explosion which started about 541 million years ago. During this 25 million year period all the body plans which currently exist in all animals appeared. Some of the animal body plans were very complex such as the phylum Chordata to which all vertebrates, including *Homo sapiens,* belong. Such body plans could not have been developed from single eukaryotic cells by applying gene modification mechanisms because of the immense increase in complexity (discussed in chapter 6). These bodies had to be designed from scratch and delivered to Earth in well advanced forms. The practical solution was to send to Earth frozen embryos of all phyla in special containers. Since all life was water based, such embryos placed in water would be in a benign environment. Water would also protect embryos against space radiation during the journey to Earth. One could envisage that the ocean's fauna were prepared in such a way that the simplest animals were sent first followed by more advanced. The Cambrian animals were well designed because they were fully adapted to the existing conditions on Earth.

The cells of animal embryos which were sent in containers had a very important feature. They had a large pool of genes, of which some were not needed and were switched off. Practically all phyla had about 20,000 genes. Several of these genes were prepared for future development and could stay dormant for millions of years. Only when the environment or living conditions changed were these genes switched on by epigenetic means. They could also be switched on using specially designed viruses.

Introns, which were introduced early on to eukaryotic cells, played an important role in the development of life. Introns enable one gene to code for several different proteins. This feature greatly enabled the production of various new proteins without the need for making new genes.

A very important event which took place more than 450 million years ago was vertebrate gene duplication[4]. This means that additional copies of the whole genome were made. Such an event could be triggered by implanting a suitable virus in the genome. That way vertebrates obtained a large amount of spare DNA material which could be modified and changed into new genes in the future. This event enabled the further development of vertebrates, leading to mammals and finally to man.

Once all phyla were well established in the seas, about 420 million years ago, some animals began to prepare to leave water. The first animals to venture onto land were the arthropods such as spiders and centipedes. Before animals moved onto land they had to be adapted to the new environment (see chapter 7). The development of these animals took place whilst still living in water. For example, the earliest four legged vertebrates, known as tetrapods, which emerged from the water about 385 million years ago, looked like fish but already had lungs and sturdy shoulders and hips capable of supporting the body's weight on land.

Further development of land based animals would have to have taken place via the genetic modification of germ cells. The most important mechanism driving the development of new classes of animals such as amphibians, reptiles, birds and mammals were the genetic modifications of DNA caused by the incorporation of viruses entering the germ cells. The incorporation of virus DNA changed the operation of organisms bringing new functions and enhancing old ones. Its effectiveness in the modification of organisms could be several orders of magnitude higher than that of mutations. Viruses are therefore the most important mechanisms for driving the development of life. Such viruses were specially engineered for specific modifications and sent to Earth at the appropriate times.

Other internal genetic variability mechanisms could have played a significant role in the development process such as transposons (Chapters 4 and 11) and the epigenetic switching of genes.

About 75 million years after the first amphibians moved from sea to land, reptiles appeared. The next leap was the development of mammals which arrived on the scene about 200 million years ago. Being warm blooded, mammals had significant changes to their body plans such as more efficient metabolism, a four chamber heart and giving birth to live young. Again these changes were driven by viruses as has been shown, for example, where the genes involved in the development of mammal's placentas originated from viruses.

The role of viruses in the shaping of mammals' bodies can be confirmed by research which has shown that adaptations in mammals have occurred three times as frequently in virus-interacting proteins compared with other proteins[5].

The genesis of man

The first step towards the development of man was the appearance, about 55 million years ago, of the order Primates. Primates had several important specific adaptations leading towards the development of man. Their hands were able to grip branches and perform various manual tasks, their shoulders had a rotating joint, their vision was stereoscopic allowing the perception of depth and most importantly they had brains larger than those of most of other animals.

The development of primates was brought about by a large number of their genes being modified by viruses. Genetic research shows that over the last 50 million years viruses have repeatedly integrated into the germ cells of our ancestors and thus became incorporated into their genome[6]. In fact, we share many of the same stretches of virus DNA with apes and monkeys. Some retroviruses* are found only in chimpanzees, Neanderthals and humans showing that they were introduced only a few million years ago. Today we carry half a million of these viral fossils, which make up eight percent of the human genome[7].

The genetic modifications of primates were successful in developing apes about 23 million years ago and the first hominins* about 6 million years ago.

Between 6 and 2 million years ago many hominin groups appeared, some more similar to the human form than others. This process produced dozens of different groups of apelike and humanlike beings, some of which used tools. It would be relatively easy to change the bodies of primates by replacing just a few genes and gene control compounds. However such modifications were not able to produce a new brain structure. In spite of the existence of many early hominin groups all of them had brains similar to chimpanzees.

About 2.5 million years ago the genus *Homo* appeared which was much more advanced than early hominin groups. The objectives of this development were to adapt the body to a vertical posture, to improve hands so they would be able to perform more precise tasks and most importantly to produce a larger and better functioning brain. Some *Homo* groups were more advanced than others. Some had a large brain, some had a more human like posture. These new forms had to prove themselves in real life conditions for many hundreds of thousands of years to be selected for further improvements. Many of them were not fit to survive, others did not show significant progress and were removed.

Genetic modifications of the genus *Homo* were not very successful and its brain did not meet the required performance. The next developmental stage took place about one million years ago when the 'progenitor' of man appeared. This progenitor has not been identified yet but it is possible that *Homo antecessor* was closely related to him. This progenitor's brain size was similar to that of man and even though he was closely related to Neanderthals, was genetically very different from other hominin groups.

We know about the existence of this 'progenitor' from research on the Neanderthal genome. It has been discovered that the Denisovans (Chapter 8) share up to 8 percent of their genome with a totally unknown species that dates back around 1 million years.

The development of man was focused on the design of his brain which is so different from the brain of animals. What is so characteristic about the development of the human brain is the presence of many new genes which do not exist in chimpanzees. According to Prof Manyuan Long there are about 1,100 new human specific genes of which about 280 are in the prefrontal cortex[8] of the brain. Most of these genes must have also existed in Neanderthals, as only 87 of the genes responsible for making proteins in cells are different between modern humans and Neanderthals.

Genetic analysis of the Neanderthal genome shows that it was 99.86 percent similar to the human genome. However, how is it possible that such a small genetic difference between Neanderthals and humans can result in such a big difference in their performance? It is because the big differences are not between the genomes but in the switching of genes. It has been discovered that about 2,200 genes which are active in humans were switched off in Neanderthals[9], or vice versa. Most of these genes are active in the brain and the immune system. So the most important genes for the functioning of the human brain were already present in our common ancestor the 'progenitor'.

The most important stage of human development was the addition of at least 87 new genes and the switching on of about 2,200 genes which were present but not active in Neanderthals, and the switching off of a similar number of genes. It was also necessary to remove some old genes which had accumulated over millions of years and which were potentially harmful to humans. Such a large number of operations could not have been performed over a short period of time using viruses and therefore a different approach was needed.

To guarantee that the right genes were switched on and the unwanted genes were switched off it was necessary to directly manipulate the genome of the embryo. The proposed scenario is that several specially designed *Homo sapiens* embryos were produced by in vitro fertilization and then were implanted into suitable surrogate mothers. This moment, recorded in our mitochondrial genes, identifies the genesis of *Homo sapiens* taking place about 200,000 years ago.

One cannot exclude the possibility that the surrogate mother was a 'progenitor' woman, therefore, early *Homo sapiens* could have lived and cross bred with them. This could explain why about 2 percent of the DNA of people with European ancestry can be traced back to Neanderthals who had a similar genome to the 'progenitor' group.

In this way the genesis of *Homo sapiens* was concluded. The main development process was carried out by using suitably engineered viruses. This hypothesis is supported by new research which has shown that an astonishing 30 percent of all protein adaptations since humans' divergence with chimpanzees have been driven by viruses[5].

The development of the *Homo sapiens* genome began 6 million years ago, after the split from chimpanzees, but accelerated very rapidly during the last half a million years. This acceleration was the result of direct modification of the human embryo. This is supported by genetic research which shows that although the human genome is different from the chimpanzee genome by about 35-50 million nucleotides, it is estimated that 7.9 percent of the changes in human DNA compared with that of the chimpanzee occurred after the split with Neanderthals which happened about 550,000 years ago[10].

Further development of *Homo sapiens*

The newly born *Homo sapiens,* although morphologically similar to modern man, still had limited abilities and some underdeveloped brain functions. This brain had to be further improved using genetic modifications. The next modification happened about 70,000 years ago when there arose a new, more socially advanced group of people (see chapter 8). These people had in them an intellectual drive to explore the world. This drive caused them to spread to South East Asia, Australia, Siberia and eventually South America.

When we look at primitive tribes currently living in New Guinea or the Amazon jungle before they encountered civilization, we have a picture of man who left Africa about 60,000 years ago. They are still mainly hunter-gatherers and able to grow some plants. They live in simple shelter-houses as closely linked small groups. They make simple decorative artifacts from feathers and animal bones. They have basic weapons such as spears and bows. They use their own simple language but being isolated they cannot communicate with other tribes. They use basic tools and kitchen utensils and they cook food. They have lived like this for tens of thousands of years without any visible development. They are not able to progress any further and their qualities are not sufficient to constitute 'modern man'. These were not people who were going to build a civilization.

Therefore, there was a need for further improvement which took place about 40,000 years ago when further genetic brain modifications increased man's intellectual abilities and artistic perceptions. These modifications were profound because they established modern man called *Homo sapiens sapiens.* They have been identified by the Bruce Lahn research group[11] and were discussed in chapter 9.

These men started developing new rituals and beliefs as well as making paintings, sculptures and jewelry. They lived in large communities and cooperated and supported each other. About 10,000 years ago they learned agriculture and became independent of hunting and gathering for their subsistence. We know that thousands of people lived in Neolithic cities and villages. However, again they reached their intellectual limitations and we have learned from archaeological discoveries that the development of these societies, such as in Catal Huyuk[12], became stagnant.

Therefore another intellectual boost was needed and this happened about 6,000 years ago. Additional genes were provided to a group of people in Central Asia which significantly increased their abilities. Genetic research carried out by the Bruce Lahn research group found that a new ASPM variant emerged in humans about 5,800 years ago (see chapter 9). Only from that time has man possessed abilities such as abstract, analytical and logical thinking, sophisticated language, imagination, learning capabilities, consciousness and introspection. Such man was ready to build the first civilizations.

This event was described in my book *We are not alone in the Universe*[13]. Since these modifications required addition and alteration of many genes I proposed that artificial insemination was used to generate the necessary genetic improvements. This procedure took place in Central Asia near the Aral Sea. From this place modern man moved out to start new civilizations in the Indus valley, Sumer and Egypt. About 5,000 years ago they moved to Europe starting the bronze age and gave birth to the Minoan civilization.

There is extensive evidence supporting this course of events. The most important support comes from genetic research showing very rapid development of man and his brain. Researchers have estimated that the development was hundreds or maybe thousands of times faster than normal evolutionary rates. More support comes from analysis of the human genome showing the huge impact of viruses on DNA.

Support also comes from archaeological discoveries. According to Michael Cremo there is overwhelming evidence that advanced human forms were already using tools several million years ago[14]. This shows that the origin of the hominin species is much older than recognized by mainstream academia and there is no clear line of progression from less developed to more advanced hominins.

Further proof comes from the extinction of all but one *Homo* group. Why did these hominin groups disappear, sometimes only after a few hundred thousand years of existence, while many species of apes lasted for more than ten million years? Having larger brains and being able to use tools, hominins were better equipped to survive than monkeys. However, for unknown reasons all these early man species faded away without a trace. This removal of unwanted species can be seen in the well documented case of Neanderthals.

Neanderthals were one of the latest prototypes of man which fell short of the qualities required for modern man. Although they had a very large brain, somehow this brain did not have the sufficient abilities needed for an intelligent man. Physically they were strong and robust and could survive in very difficult climatic conditions, but they were in the cul-de-sac of man's development leading to nowhere. They had to be removed because they could interbreed with *Homo sapiens* injecting back the wrong genes into human development. Their extinction was not as dramatic as seen by some scientists, but was simply the result of reduced procreation abilities. As a result of some gene modifications the population of 70,000, about 200,000 years ago, was reduced to just a few thousand about 40,000 years ago and that way Neanderthals faded from the scene.

The course of man's development shows that the designers did not have ready-made design plans of the *Homo sapiens* brain, but as in many development projects they tried several models before reaching an acceptable solution.

So looking at the history of the arising of intelligent life on Earth we have to conclude that it could not have developed by chance. The only solution is that it was the hand of the designer which was responsible.

Who is the designer?

The most important question in the new genesis paradigm is "Who is the designer?" This question has not been answered although many hypotheses have been proposed. A large part of the population believes that the designer was God. But God, by definition, is omnipresent and omnipotent, not constrained by the laws of physics or the available energy and time.

The question is, whether events leading to the preparation of Earth and then to the arising of life would have required, from a human point of view, supernatural powers.

The most accomplished event was the bringing of water to Earth by directing a suitable body on a collision course with the planet and changing the Earth's orbit. This exercise required cosmic planning and a huge amount of energy. Would we have needed God to produce such an event? Not necessarily. Intelligent beings, having at their disposal appropriate technologies such as powerful nuclear fusion engines or controlled nuclear explosions, could change the orbit of a large body and direct it to Earth (Appendix I). We have to take into account that they could have at their disposal much more powerful engines based on technologies unknown to us. It would take a long time, maybe many thousands of years, but it could be done. Even now, with our primitive technology, we are able to consider changing the trajectory of large asteroids on a collision course with Earth.

When we come to the origin of life itself, intelligent beings could have designed and built a living cell with all its necessary molecular machines. We already have some technology to do this in our laboratories and synthetic DNA has already been implanted in cells.

The fact that it took about 4 billion years from the onset of life to the arrival of human beings indicates that normal geophysical and biological development processes took place and no supernatural interference was needed. So, these examples show that God does not need to be the designer.

If it is not God, then who? We have to consider that some very intelligent beings may have existed in the Universe for billions of years. This sounds like a very long time, but only when compared with the duration of human life. So we have to ask a question: are these extraterrestrial beings similar to humans? After analyzing the history of the arising of life on Earth it appears that these beings did not have full knowledge of the operation of our biological system prior to seeding life on Earth. If they had, our development could have progressed much faster. For example, more efficient photosynthesis cells could have been introduced right from the beginning, generating oxygen more quickly. Waiting 3 billion years for the Cambrian explosion could be explained by the necessity of waiting for Earth to enter a more stable period in her life. This would also indicate the limited powers of these beings.

The development of man follows an erratic path. It was not man's body but the design of his brain and properties of his mind which were modified several times before arriving at the present version, which is far from perfect. If these beings had any prior knowledge of the operation of this type of brain I am sure that they would have come up with a better solution than the design currently used for *Homo sapiens*. One could draw the conclusion that it is therefore most unlikely that the functioning of the intelligent beings' body and brain is based on biochemical principles similar to ours. If their bodies and minds are so different from ours, they should have very different abilities, and their life cycle could be very different from ours, lasting maybe millions of our years.

Obviously we could speculate "ad infinitum" who the designers are, but this would not serve any purpose. We have to accept that there is much more to the Universe than we can imagine and comprehend. Unfortunately our mind is fettered by evolutionary theory and our thinking and ideas are determined by it, impeding our ability to look beyond. We believe that we are very clever and this is why we search for aliens which are just like us, maybe slightly more advanced, because we cannot imagine anything else. I believe that the extraterrestrial beings responsible for our world are so different from us that they are beyond our comprehension. To us they are like gods but not necessarily God.

Support for extraterrestrial involvement comes from the work of Crick and Orgel[15] who put forward a hypothesis of directed panspermia. They proposed that the first organisms arrived from space and were sent on purpose to seed life on Earth. Hoyle and Wickramasinghe[16] confirmed that life could have come from the cosmos and further research shows that bacteria and even higher organisms could have survived a journey through space.

Further support for extraterrestrial design comes from the analysis of our genetic code. Work performed by Shcherback and Makukov shows unexplained properties of DNA codes which would require some intelligent approach that cannot be accounted for by Darwinian evolution. They say that "simple arrangement of the code reveals an ensemble of arithmetical and ideographical patterns of symbolic language. Accurate and systematic, these underlying patterns appear as a product of precision logic and nontrivial computing rather than of stochastic processes." [17]

The purpose of the experiment

One of the key questions is, "Why have intelligent beings embarked on this experiment?" I am sure that they did not try to provide some amusement for themselves, although looking at the last 5,000 years of human history is so entertaining for us so it must be quite amusing for them. I proposed a hypothesis in my earlier book that their purpose was to develop a new type of conscious being. I wrote, "The belief that the intelligent beings advanced the development of man's consciousness leads us to a hypothesis that mankind has some objectives which could be fulfilled only when it achieves higher consciousness. Such an aim would include achieving a level of consciousness comparable with the level which had been reached by the intelligent beings." [18]

I proposed that intelligent beings, being aware of their own decline, decided that cosmic intelligence and consciousness should be taken over and propagated by human kind. Therefore, they embarked on a long term project with the aim of developing a new type of being which would be able to carry out their task. To do this they had to prepare a new unique habitat – Earth, and a new type of life-form. When man eventually attains higher consciousness he will take over the function of intelligent beings in the Universe. Some may say that it is precarious to try to guess what intelligent beings have in mind, but I feel that my guess is justified by the information I have obtained.

The case of human suffering

The history of the human race consists of endless wars, killings, torture, brutality and hunger. These sufferings are caused by human beings themselves and are the products of the human mind. But we have another area of suffering caused by the imperfect design of our bodies: pain of birth, illness, diseases, accidents and pain of death. Therefore one could say that for both kinds of suffering, intelligent beings are somehow responsible. This raises the question: Why do intelligent beings allow so much suffering to take place? While religion has a ready answer: "God wants to test us", finding justification in the works of intelligent beings is much more difficult.

The first kind of suffering results from the imperfect human mind. Study of the human mind, especially recent discoveries[19] shows how fickle humans are. Should the human mind be better designed?

One could reply that the destiny of man is to be responsible for all aspects of his life. Man is not another pre-programmed animal fulfilling certain basic functions of nature. Humans are apart from the rest of the animal world by having free will and making free choices. This is the only way for man to develop a higher consciousness. We do not know if a more constrained and rational mind could lead to conscious man. We know that intelligent beings were not happy with the original design of the brain and kept improving it even as recently as 6,000 years ago. So we have to assume that our brain is far from perfect but it is the best that it can be. I think that the brain has so much built-in flexibility and potential that people are able to change themselves. But this change could be the result of conscious work. Man's further improvement may come only from within.

Another kind of suffering is caused by illness and disease which could also be traced to the intelligent designer. However, should we blame the designer for all our medical problems?

Man, as a biological organism, is not treated preferentially by intelligent beings. He is not treated differently from the other animals. Intelligent beings do not normally interfere in the lives of individual people, they do not protect people from suffering. We know that many diseases and parasites affect animals as well as humans so they are not specially 'designed' for man. Many illnesses are caused by genetic mutations which are beyond anybody's control.

It is impossible to know if the biological world could be designed without diseases and parasites. Many of these ailments could result not from specific designs but from the workings of directed evolution which is free to generate new life forms. It is futile to try to speculate how life on Earth could be designed better but we know that we could make it better.

Many causes of suffering could be removed by man's actions. Consider how much medical progress has been made during the last hundred years. In the last 50 years we have learnt so much about biochemistry, molecular biology and genetics, which will eventually help to eradicate many diseases. Now it is man's responsibility to combat disease. If sufficient resources were allocated to medical research much faster progress would be achieved. So the eradication of known illnesses and diseases is in human hands. Progress has been made but we are impatient because we want to see improvements in our lifetime, but from the intelligent beings perspective, for whom a few thousand years is an insignificant period in Earth's history, the human race is moving forward.

Comments

The sequence of major events starting with the preparation of Earth and ending with the arising of man is so improbable to have happened by chance that the only plausible answer is that extraterrestrial beings were responsible for this chain of events. The purpose of this experiment is that eventually the human race will become the carrier of intelligence and consciousness in the Universe. We might be a long way from achieving this objective, but taking into account that our civilization has so far existed for less than 6,000 years and accomplished so much we can expect that maybe in a few thousand years we may be able to fulfill our goal.

The life of the human race is not easy and is plagued by many illnesses and diseases, however most of our sufferings are manmade. The solution to this predicament is the development of man. The development of the human mind and the achievement of a higher level of consciousness will remove not only the sufferings created by man but also will make it possible to remove many illnesses and diseases.

Looking at the present state of the world we might be under the impression that things are getting worse. However, if we compare our present situation with the situation a hundred years ago we can see tremendous progress in practically all aspects of our lives. I believe that the human race is moving on the right path towards achieving its objectives.

CHAPTER 14

In search of....

Where are they?

Many scientists believe that there must be thousands or maybe even millions of extraterrestrial civilizations in the Universe. They must be there because we have been told by evolutionists that liquid water and correct temperature are the key conditions required to breed another intelligent race. I am being a little bit facetious, but I cannot help it because this is what many academics believe.

The search for aliens is becoming more and more popular. About 60 years ago the SETI[1] program was started which uses radio telescopes to search for signals from other stars and galaxies. The search is based on an assumption that any extraterrestrial civilization which wants to communicate with other civilizations will use radio or microwave signals similar to ours. So far, this program has not provided any evidence for the existence of extraterrestrial civilizations. In spite of this setback our fascination with extraterrestrial life has not diminished. This is reflected by an initiative[2] instigated in 2015 by physicist Stephen Hawking calling for a more intensive search for aliens. So, we must believe in the existence of aliens otherwise why would we keep spending many millions of dollars, employ hundreds of scientists and use very expensive equipment, which could perform other important scientific tasks, on finding them?

Regardless of how much more money and time we may spend, it is most unlikely that we will establish contact with other civilizations. The search is based on false assumptions that aliens are only slightly more advanced than we are and use very similar communication technologies. If we look at our own civilization, would we have been able to contact it even two hundred years ago? Will we be able to contact it in two hundred years time? The probability, that other extraterrestrial civilizations are on a similar development level of science and technology to ours, is practically zero. They could be even billions of years ahead of us, and we know how much progress we have made in the past hundred years alone.

So where are they? I believe that they are here on Earth... Hold on. This is not as outrageous as you may think. When I proposed this hypothesis my first reaction was "this is impossible", but after investigating some unusual events I came to the conclusion "why not?" Looking at the history of our genesis you would agree that intelligent beings have invested a lot of time and effort in us and therefore have a vital interest in our development and in the events on Earth. Even assuming that they could be located somewhere in our galaxy, they would not be able to travel distances of many thousands of light years very easily. They would have to be close to our solar system to take part in the design of life and other activities. In the past, life on Earth was developing in large steps separated by hundreds of millions of years of stasis, so their permanent presence on Earth was not necessary. However, the development of man's brain needed their close involvement since we know that several versions of *Homo* were prepared. The arising of modern and conscious man about 6000 years ago was the most important event in their experiment, and therefore they had to supervise his development. The easiest way would have been to do this from close quarters. Therefore, why might they not be on Earth, which is a very comfortable planet to live on?

The main scientific objection to this idea is that we are not able to detect them. At present Earth is covered by a very dense net of radar signals and satellite scanners, and is guarded by millions of CCTV cameras. Not to mention seven billion people watching the space around them in all possible corners of the Earth. Our surveillance technology is very advanced, close to the theoretical limits of physics. Optical devices placed on satellites are able to spot a single man on the surface of Earth and detect a hundred meter diameter rock several million kilometers away from us. So how can they avoid detection?

It is possible that the alien's bodies are not built from the same biological materials as we are. Or they could have technology which means their mode of transport does not reflect electromagnetic radiation, in a similar way to our stealth aircraft technology. If their materials also do not absorb electromagnetic radiation their vehicles could be perfectly transparent and therefore not visible.

The second major objection to the existence of aliens on Earth is that they have not contacted us yet. The simple answer is that they do not want to contact us. If they are very advanced, any direct contact could cause more damage to us than bring benefits. They might not want to contact us directly for the same reason that we want to protect tribes in the Amazon forest from any contact with civilization. A group of people being in contact with aliens would have a huge advantage over the remaining population and most likely would use this to their own benefit. Another reason for not contacting us is that they do not want to upset our major religions which play such an important role in our society.

The third reason why we do not believe in their presence on Earth is that every year we are flooded with many thousands of reports of sightings of UFOs, meetings with aliens, etc. Most of them are products of our imagination, or our psychological needs. However within these thousands of sightings may be a few genuine events originated by extraterrestrials. Investigation of such sightings after the event is practically an impossible task, because we have to rely on the integrity and reliability of the observers. As a result, all these observations are ignored.

I believe that intelligent beings have contacted us, but on their own terms, without causing any harmful consequences. They contact us through intermediaries, which I discussed in my first book[3], or by using special religious events which comply with prevailing religious beliefs. The display in the sky in 1917 can be counted as such a contact. This is the best documented contact with intelligent beings and therefore it is worth providing a more detailed description.

Display in the sky

The display in the sky took place on 13 October 1917 in Fatima, Portugal after the apparitions[4]. During the apparitions three children saw a figure of a lady who passed some messages to mankind. The display was not part of the apparitions, but a public show observed by about 50,000 people. What is special about this event is that it was announced about 3 months in advance by one of the seers, which is why so many people attended the display. This was a unique event in the history of any "paranormal" occurrences which was predicted well in advance. The display did not have any religious character. It was more of a light show arranged to provide people with some tangible evidence that the Fatima apparitions were genuine and not a product of the children's imaginations.

The show started about noon when a light disc appeared in the sky at the position of the Sun. The diameter of the disc was similar to the diameter of the Sun, which is about 30 angular minutes. Witnesses confirmed that before the display, the sky was overcast, and then suddenly, a hole appeared in the clouds in which the disc was visible. The luminous disc did not shine brightly and was not blinding. It was possible to look at it without squinting. The disc had a distinct edge and was of a silver-blue color. The color was clear and rich similar to the opalescence of a pearl. Witnesses with a technical background described it as a disc of opaque or frosted glass, illuminated from behind, with a rainbow of iridescence on its periphery. Some of the witnesses claimed that the colors of the disc were lively and strong, and changing from white, red and blue to orange, yellow or green. It looked as though the periphery of the disc was more colorful and iridescent than the centre and changed cyclically, according to some repeating pattern. The color changes were not random and rapid, but slow and regular. The disc sometimes looked as if it was surrounded with an aureole of flames, at other times by yellow or purple rays. At one moment the flames disappeared and the disc became a dull silver-blue color again. One of the witnesses noticed that the clouds, moving at that time from the west to the east, were not obstructed from view by the disc but it looked as if they passed behind it.

The disc did not stay in one place, but moved in the sky. All witnesses agreed that it revolved on its axis. Some people noticed that it whirled very quickly. The rotation was not uniform, but repeated itself three times after intervals. There are descriptions comparing the disc to a spinning wheel of fireworks, with flashes of sparks of light on its border similar to the Catherine wheel. Apart from the whirling, witnesses noticed other movements. They describe that the disc "danced", "jumped" or "trembled" through the diaphanous clouds. These irregular and arrhythmic movements caused sudden and rather accidental changes of the disc's position.

The most dramatic scene happened when the disc fell down towards the earth. The whirling disc appeared to approach the earth, and its movement was very elaborate. It looked as though the disc was spinning, descending in slow zigzags, or moving down in a spiral. The movement reminded some witnesses of the fall of a dry leaf from trees in the autumn. When the disc descended so low that it looked as if it was touching the tips of pine trees, a panic seized some people, but then the disc started climbing back up at a very slow, snail like pace, all the way to its initial position. The entire display lasted about 10 minutes.

There are no photographs of the display itself, but my calculations show that the films used at that time were not sensitive enough to record low levels of laser light used for the display. However, there are photographs of watching crowds.

Because so many people witnessed the show, extensive written evidence exists. Several descriptions were provided by atheist journalists hostile to the Church who could not be accused of any religious bias. There are also written statements of many professionals such as doctors, lawyers, teachers etc. and all their observations are very similar.

The Catholic Church was initially very hostile to the apparitions and it took several years, and pressure from the public, for the Vatican to accept the religious apparitions in Fatima. But they did not include the display in the sky.

After analyzing the witnesses' statements I came to the conclusion that this event could only have been organized by intelligent beings. While at this moment it would be feasible to organize a similar laser display, a hundred years ago it would have been impossible. For me, the existence of the predictions is of immense importance because nothing like this has ever happened in the history of paranormal or unexplained events. However, this fact did not generate the same excitement outside Portugal. It was simply disregarded.

The events in Fatima are considered by the scientific world in exactly the same way as many other religious apparitions: they are simply ignored. Information about the display does not rouse any interest because it is treated as a Catholic Church event.

Lights

The most common evidence of the presence of extraterrestrials comes in the form of lights in the sky. All different kinds of lights have been observed over more than 60 years. There are many reports from very reliable sources, such as pilots, of unexplained lights following aircrafts. Between 1952 and 1969 these sightings were listed in the Blue Book [5] and are currently also recorded on numerous UFO websites. When events in the Blue Book were analyzed most of them were rejected as being of man-made or of natural origins. However about 5 percent of them, even after very careful analysis by experienced scientists escaped the usual classifications and were left unexplained.

In practice it is difficult to investigate the nature of lights, especially when observed from a long distance over a short period of time, therefore it is impossible to prove their origin. However there are two sightings of lights which are very different from any other reported observations. In 1983 in Warsaw five slow moving lights were viewed over several minutes from a distance of only about 60 m, which is a very unique happening (Appendix 2). During the same evening another 20 sightings in Poland were also observed, but these lights were seen from greater distances. It was confirmed that the observed moving objects were not detected by military radars[6].

Another important light observation took place in July 2012 in Warsaw where two lights were circling in the sky at a height of about 400 meters. These lights moved on identical circular trajectories forming a figure of 8 and their movements were so perfectly synchronized that every 20 seconds they crossed at the same point in the sky. The lights were observed for about 30 minutes and their position was not affected by strong wind. This type of display is also quite unique as it has never been observed before. A more detailed description of this event is given in Appendix 2.

These observations, although witnessed by only two people, could be treated as proof of the activities of intelligent beings on Earth. The structure and movement of these lights excludes any natural as well as any human origin.

Comments

The evidence described above belongs to two different categories. Fatima, having been witnessed by 50,000 people, is without any doubt the most unique and best documented extraterrestrial activity. While the Fatima display would indicate that intelligent beings were present within a few hundred kilometers from the location of events, the observed lights do not provide proof of their physical presence. The lights could easily have been part of their remotely controlled surveillance vehicles transmitting signals back to the control centre.

Being such a special event, one would assume that the Fatima display should generate wide interest, but this is not the case. Is it because the Fatima display is treated as a religious event? I think that there is a deep psychological reason for this rejection.

We all see ourselves as being open minded by accepting novelties such as new scientific discoveries, new religions and new political ideas. But there is a limit to how much we can accept. We want a secure life, nothing too upsetting, nothing too dramatic and anything which can destroy our equilibrium is rejected by our subconscious mind, whose objective is to protect our mental stability. As a result of this there is a strong unconscious resistance to accepting evidence for the existence of intelligent beings on Earth. We are prepared to accept the hypotheses of their activities somewhere in the Universe, thousands of light-years away from us, however accepting the existence of intelligent beings dwelling so close to us makes us very uncomfortable. When we look at the Fatima display, these events were very recent and close to us. They are well proven and well documented, witnessed by so many people that it is difficult to suspect any conspiracy or collusion. However, this evidence makes many people very uneasy to the extent that they refuse to listen to the arguments. Whilst one could understand that religious people are not interested in a new interpretation of the Fatima events, what is most surprising is that atheists are far more hostile to this idea.

I believe that these three observations are adequate proof of evidence that intelligent beings function somewhere on, or close to our planet. But most people will never accept this because they are waiting for 'green men' to turn up on their doorsteps.

Epilogue

Together we have covered a long journey from the creation of a habitable planet to the arising of conscious man. All the milestones that have been discussed along this journey show well planned and organized activities spanning more than 4 billion years. I believe that the evidence that has been presented in this book should be sufficient to swing the reader's opinion about these events. However we are slaves to our beliefs and it is known that even the most reliable, logical evidence is sometimes not enough to change the minds of strong believers. So the reader should start their own journey to look for evidence and by applying logical arguments should be able to reach their own conclusions.

I believe that at some point in the future biological science will become so suffocated by the restrictions placed on it by evolutionary theories that it will be forced to throw off this gag that prevents the truth from prevailing. However this will be a very long process as the opponents of intelligent design will not change their minds even when faced with irrefutable proofs.

What would be the consequences when the truth triumphs? Earth will be back at the centre of our Universe. Maybe not in the physical sense, because who knows where this centre is, but in the philosophical and purposeful centre because Earth was selected for man. We would become one of the most important creatures in the Cosmos, but before we reach this position we would have to work on ourselves. Only then, when we achieve higher consciousness, will we fulfill our destiny.

Life is a miracle which at this moment is beyond scientific explanation. As Albert Einstein said, "There are only two ways to live your life. One is as though nothing is a miracle. The other is as though everything is a miracle." After reading this book the reader would have to make up their own mind.

APPENDIX 1

Calculations for the delivery of water and the change in Earth's orbit

1. Energy needed to deliver water to Earth

Volume of water to be delivered to Earth:

1.4 billion cubic kilometers = 1.4×10^{18} m^3 = 1.4×10^{21} liters.
Let's assume that water was delivered by an asteroid with a density of 2.6 g/cm^3 and made up of 20% water.
Total mass of asteroid: m = 7×10^{21} kg = 7×10^{18} tons
The asteroid's mass is about 0.0012% of Earth's mass.
Volume of asteroid: V = 2.7×10^{9} km^3.
Diameter of asteroid: D = 1,720 km

Let us assume that this asteroid originated from the Oort Cloud and its distance from the Sun was R_2 = 10,000 AU.
The orbital speed of the asteroid was:

$$V_{orb} = (GM/R_2)^{1/2} = 297.4 \text{ m/sec} \qquad \ldots\ldots (1)$$

where the Sun's standard gravitational parameter GM = 1.327×10^{20} m^3/sec.

The asteroid could have been brought to Earth using, for example, Hohmann[1] transfer.

During this process a body travelling on its original circular orbit is slowed down and put onto a transfer orbit. Figure A1-1. This transfer orbit is calculated in such a way that its trajectory would cross the Earth's orbit.

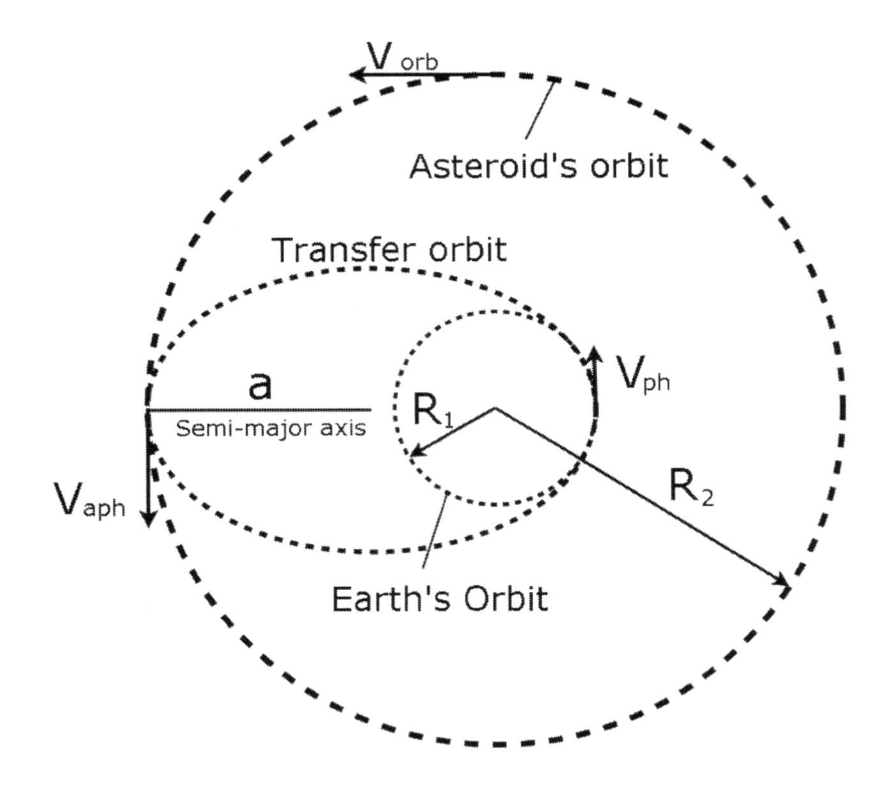

Figure A1-1. Schematic diagram showing Hohmann transfer orbit.

The semi-major axis of the transfer orbit is

$$a = (R_1+R_2)/2 = 5{,}000 \text{ AU}$$

where R_1 = 1 AU is the distance of Earth from the Sun

The period of the transfer orbit is calculated as

$$P_T = 2\pi\times(a^3/GM)^{1/2} = 1.12\times10^{13}\text{ sec} = 3.3\times10^5\text{ years} \quad .... (2)$$

For an asteroid to change its circular orbit onto the transfer orbit its speed must be reduced to the aphelion speed, V_{aph} of the transfer orbit.

The change of speed $\Delta V_2 = 293.2$ m/sec.

The asteroid's kinetic energy must be changed by

$$K_E = m(V_{orb}^{\,2} - V_{aph}^{\,2})/2 = 3.1\times10^{26}\text{ J} \quad ... (3)$$

where $V_{aph} = 4.2$ m/sec.

To reduce the asteroid's speed over 10^{12} seconds, or 30,000 years, about 300 TW of power would have been needed.

However, if the whole process was spread over a much longer period of time or the gravitational forces of other planets were used, much less power would have been needed.

It cannot be excluded that this speed reduction could have occurred as the result of a collision with another body.

When the asteroid reached Earth's orbit it would have been travelling with a perihelion speed of $V_{ph} = 42$ km/sec. Since Earth was moving with a speed of about 29.8 km/sec, the relative speed of impact would have been about 12.2 km/sec. The final speed after the collision would have been 29.813 km/sec.

2. Effect of asteroid impact on Earth

During impact, the dissipated kinetic energy of the asteroid is calculated from:

Initial kinetic energy is 26.707×10^{32} J
Final kinetic energy is 26.695×10^{32} J
The change in energy is 1.2×10^{30} J = 2.86×10^{29} calories

Assuming that upon impact all of the asteroid's kinetic energy is changed into heat.

Earth's mass: $m_E = 5.98\times10^{24}$ kg = 5.98×10^{27} g,
Earth's material specific heat = 0.3 cal/g
To warm Earth by 1 degree requires 1.8×10^{27} cal

Earth would be warmed by 159°C if 100% of the kinetic energy was changed into heat. These calculations are for purely non elastic collisions and do not consider the loss of kinetic energy as a result of some mass being ejected from Earth upon impact. Without the greenhouse effect the Earth's average temperature at the time of impact would have been about minus 17°C, but in reality, probably much lower because the Sun's radiation was about 30 percent lower than it is at present. Therefore, the Earth's temperature after the impact would have been below the boiling point of water.

3. Timing accuracy of the body to be able to hit Earth when travelling from the Kuiper Belt.

Let us assume that the asteroid came from the Kuiper Belt which is at a distance of 50 AU from Earth.

Using equation (2), the period transfer orbit is

$P_T = 4.08 \times 10^9$ sec = 129.3 years
Total travelling time of the asteroid is 2.04×10^9 sec
Earth's orbital speed: 29.8 km/sec.
Earth's diameter: D = 12,756 km
Earth crosses a given point on its orbit in 425 seconds or about 7 minutes.

Therefore the accuracy of the timing is given by:

(Earth's crossing time)/(total travelling time) = $425/(2.04 \times 10^9) = 2.08 \times 10^{-7}$.

The timing accuracy must be better than 0.2 parts per million or 0.00002%

4. Change of the Earth's orbit

Let's assume that we want to change the Earth's orbit radius by 10 million kilometers and

> $R_1 = 160 \times 10^6$ km was the initial distance between Earth and the Sun, and $R_2 = 150 \times 10^6$ km is the current distance between Earth and the Sun.

Since Earth's present speed is 29.8 km/sec the speed on its initial orbit is given by:

$$V_{160} = V_{150}\,(150/160)^{1/2} = 28.85 \text{ km/sec}$$

Earth's orbit could have been changed using Hohmann transfer.

The speed of Earth on the initial orbit was 28.85 km/sec. The aphelion speed of the elliptical transfer orbit is 28.37 km/sec therefore the Earth's speed must be reduced by 480 km/sec. The period of the transfer orbit is 3.32×10^7 sec.

Earth's speed could be reduced by collision with a suitable body.

Since we can calculate that a body originating in the Kuiper belt would reach a speed of about 41.65 km/sec when crossing the Earth's orbit, we can calculate the mass of a body m_1 whose impact would reduce the Earth's speed to 28.37 km/sec from:

$$m_1 = m_E\,((V_2 - V_F)/(V_1 + V_F) = 0.00685\, m_E = 4.11 \times 10^{22} \text{ kg}$$

where m_E is the Earth's mass, $V_2 = 41.65$ km/sec, $V_F = 28.37$ km/sec and $V_1 = 28.85$ km/sec.

This means that a body with a mass of less than one hundredth that of Earth would be sufficient to move Earth to the transfer orbit.

When Earth reached the new transfer orbit its perihelion speed was 30.26 km/sec therefore its speed must have been reduced again by 480 km/sec to reach its final speed of 29.78 km/sec. The second speed reduction could have been achieved by a similar impact.

These calculations do not intend to prove that the change to the Earth's orbit occurred as presented. Rather they show that the shifting of Earth's orbit is feasible.

It is possible that the body which caused this change also carried water. Since the mass of this body is much larger than the one proposed in section (1), its water contents would only need to be 3.4% of its body mass to deliver enough water to Earth.

5. Temperature change as a function of distance from the Sun.

To calculate Earth's temperature changes I use the following equation[2]

$$T = 280 \ [1 - A]^{1/4} / a^{1/2} \ [°K] \qquad ....[4]$$

Where A is the Earth's albedo and **a** is the distance between the Sun and the Earth in AU (astronomical units).

This equation does not take into account the greenhouse effect.

For: $a = 1, A = 0.3$
$T = 256°K = -17°C.$

The greenhouse effect increases the Earth's surface temperature by 33°C.

If we increase the distance between the Sun and Earth by 10 million kilometers, from 150Mkm to 160Mkm we can calculate the Earth's temperature from equation [4]:

$a = 160/150 = 1.066, \ a^{1/2} = 1.033, \ T = 248°K$
$\Delta T = 248 - 256 = -8°K$

The Earth's surface temperature at a distance of 160 Mkm was 8°C lower than at a distance of 150 Mkm.

We could estimate that without the greenhouse effect, which should be constant for the same environmental conditions, the Earth's surface temperature changes by about 8°C when the distance from the Sun is changed by 10 million kilometers.

APPENDIX 2

Description of lights in Warsaw

Lights in 1983

A very special and unusual event was witnessed by Witold Koczynski, a 47 year old electronic engineer, and by Klaudia Allison, an 11 year old school girl, who on December 2nd 1983 between 19:30 and 20:00 were walking their dogs in a built up area of Warsaw, Poland. It was a dry, cold evening with clear skies and good visibility. Witold at first noticed 5 lights looking like a pentagonal chandelier suspended several hundred meters in the air in a westerly direction. These visible weak lights glowed with a constant pink-red intensity but did not move for several minutes. Suddenly they descended with high speed reaching a 60 degree inclination. The lights formed a square and began flashing white and violet colors. It took them just a few seconds to come close to the roof tops and then they disappeared behind a building (A) which was about 25 m high (Fig. A2-1). After about 15-20 seconds the balls of light were seen from position (1) between the buildings (A) and (B), from a distance of about 60 m, moving horizontally in a south-easterly direction at a slow pace. The balls of light were unusually beautiful being perfectly round and having light green, silvery colors.

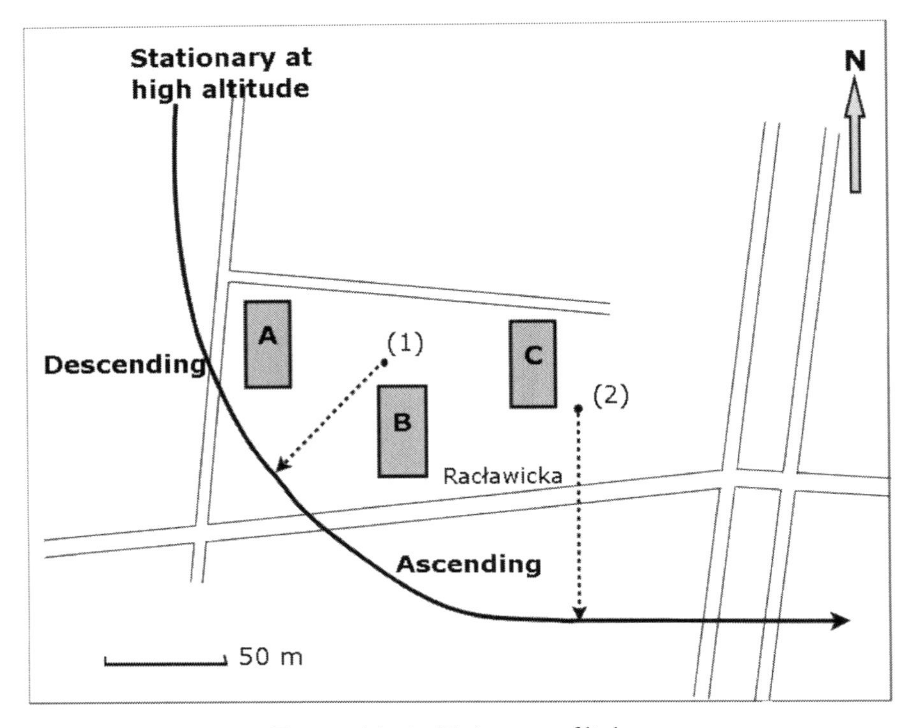

Figure A2 -1. Trajectory of lights.

As the lights were behind building (B) Witold moved to the front of the building (C), position (2), where he could see the lights flying at about 30 m above the ground forming a square, about 5 m x 5 m, with two balls in front, two balls at the back and the fifth ball was moving behind at a distance of about 10 m. Being close, about 100 m from the lights, Witold could see many details and he noticed a weak beam of white light projecting downwards from a source in the middle of the square. This light beam was about 10 m long, inclined from the vertical to the rear by about 20 degrees, and diverged in a cone shape of about 20 degrees.

The lights moved slowly across the road and started rising steeply in an easterly direction keeping the same formation and then disappeared after a few minutes. What is interesting is that during the observation, which lasted about 20 minutes, no sound was heard, and since it was a quiet evening the witnesses would have been able to hear even the slightest noise.

What is important about this sighting is that the observers were very close to the object, therefore they could judge its size, speed and light intensity. They noticed that the lights moved together all the way, but there was no visible structure holding them together. The most important fact is that the main witness, Witold Koczynski having an engineering education was familiar with assessing technical devices and was able to notice even the smallest details. He was never previously interested in UFOs and was always very skeptical about any paranormal events. The fact that I knew him well since our University years together is not without influence on the trust I have in him.

Looking at the description of this event in 'UFO' magazine[1] it came to my attention that on the same day about 20 other sightings of lights in Poland were reported. All these sightings took place between 19:20 and 20:17 hours in locations several hundred kilometers apart, however nobody had such a close encounter with the lights as in Warsaw.

Lights like this would not pass unnoticed by the Polish Air Defenses who filed several reports[1] on unidentified lights that same night. Some of these lights moved with supersonic speeds, some at low altitudes and some as high as 10 km. What all these reports had in common is that military radar did not detect any observed objects and no engine noise was heard. It was confirmed that no flights of any aircrafts took place in these observation areas at the reported times. Reports confirm that similar lights were seen at that time in the USSR and Germany (GDR).

These lights could not have been of natural origin as they moved in a controlled and organized way. They could not have been manmade because they were not detected by any radar and they moved without generating any noise. They moved in tandem but there was no visible link between them. So the overwhelming conclusion is that they were of extraterrestrial origin. But this display also shows that extraterrestrial devices cannot be detected by our surveillance equipment.

Lights in 2012

I have decided to include lights which I observed in Warsaw in July 2012. No one else witnessed these lights, but what is special about them is a very unusual and precise movement which was observed for about half an hour.

At the time of the display I was staying in a top, 3rd floor flat, south of the centre of Warsaw (52°12’23”N, 21°00’53”E). On 3rd of July at about 23:00 I heard the sound of distant thunder. When I approached the balcony door to close it, I stopped and looked up at the lightning. Just by chance, right above me in the sky, I noticed a moving spot of white light. At first I thought that this might be the reflection of light from the low clouds originating from car headlights, but when a denser cloud passed underneath, the spot of light disappeared. This meant that the light could not originate from the ground but from above the low clouds. The layer of clouds was quite thick and uniform, and no stars were visible through it. The clouds moved with a speed of about 5-10 m/sec pushed by a strong westerly wind. The base of the clouds was at a height of approximately 400 m (1200 ft) and was slightly brightened by the street lights. The spot of light was not very bright but clearly visible.

The spot of light was moving with a constant speed in an anticlockwise direction on a circular path which diameter subtended angle was about 20° (Figure A2-2). It took the light spot about 20 seconds to complete one revolution. The centre of the circle was about 70° above the horizon. The diameter of the spot of light was about 3-4 times larger than the Moon (subtended angle of 1.5 - 2°), but it did not have sharp edges. Its intensity was not uniform, being brightest in the centre, and its edges were blurred like car headlights in a dense fog. Whilst travelling along its circular path the spot of light changed its shape depending on the structure of the clouds it was moving through. Where the passing cloud had a uniform density, the shape of the spot was round. However where the clouds density was not uniform, the spot could be oval, jagged, or completely irregular with narrow, spiky light rays emanating in radial directions from it. Where the cloud was very dense the light could completely disappear. The unobstructed source of light was never visible.

I then noticed a second spot of light, very similar to the first, which was also moving along a circular path which was touching the first spot’s circular path. The second spot’s circular path had the same diameter as the first and was positioned to the right and above the first path. Its centre was about 80° above the horizon meaning that the centers of both circles were not in a horizontal line. The planes of the two trajectories were inclined to the horizontal plane by 20° and 10° respectively. Since I saw two perfect circular trajectories, their planes must have been perpendicular to the lines from their centers to my eyes.

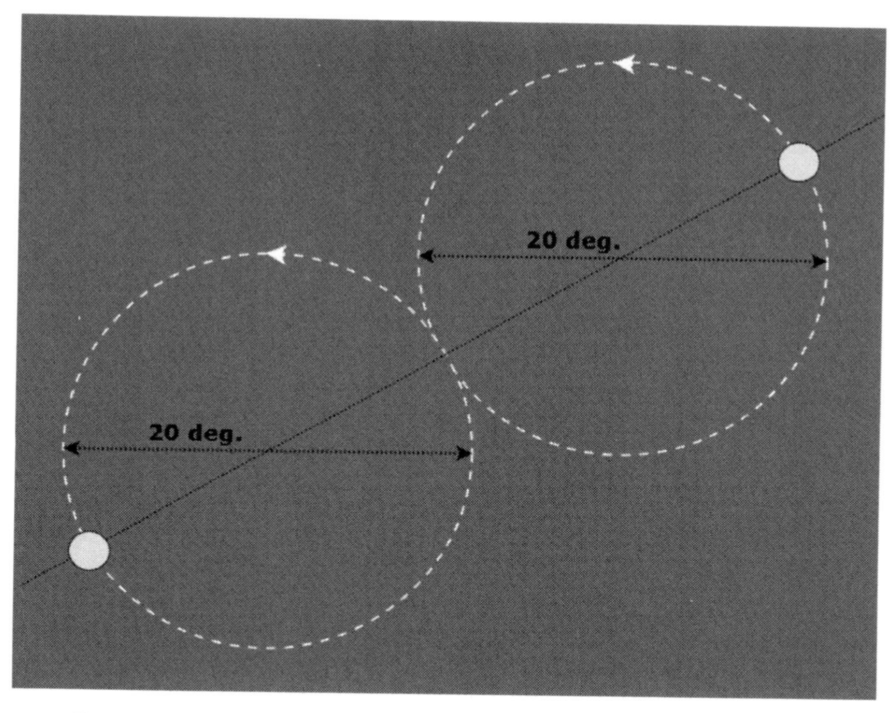

Figure A2- 2. Geometry of the circular motion of the spots of light.

The second spot was also moving along its path in an anticlockwise direction, with the same speed as the first spot. Both lights always met at the same point where the two circular paths touched each other. At this point the lights were fully overlapped in such a way that only one spot was visible. The movements of both spots were perfectly synchronized at all times during the observation. The shape of the second light spot changed in a similar way to the shape of the first light spot, but in a given moment of time the shapes of the spots were not identical because the lights were passing through different cloud areas. When the clouds were uniform, both lights had a similar circular shape.

The most likely source of the light would have to be in the shape of a sphere radiating rays in all directions. Therefore the brightness of the spot of light when viewed from underneath the cloud was not uniform. After analyzing the shapes of the spots I reached the conclusion that the sources of the lights must have been located in the lower layer of the clouds, otherwise the light would not be visible if positioned above the thick clouds.

Another observation which confirms the assumption that the light source was placed in the lower part of the clouds was the propagation of irregular light rays far from the centre of the spots in radial directions. Any light source placed above the clouds would not be able to generate transverse rays visible from underneath the clouds.

Assuming that the lights were about 400 meters above the ground, the two spots of light had a diameter of about 14 meters and moved with a constant speed of 22 m/sec. The diameters of both the trajectories were about 140 meters. The spots moved with high precision because the intersection point during the observation was always in the same place in the sky. The trajectories overlapped so perfectly that only one spot was visible at the intersection point.

One would expect that the light of the moving spots was generated by some sort of source which must have had a certain mass and dimensions. Therefore, gravitational, centrifugal and wind forces should act on the moving light source. However, in spite of the strong wind blowing, the position of the light trajectories did not change for at least 30 minutes. The speed of movement was constant in spite of varying wind pressure because the lights met at exactly the same point in the sky.

The centrifugal force acting on the light source would have had to be compensated by propulsion to keep the source on a perfect circular trajectory. In my opinion we have to exclude the existence of any rigid structure because of the large diameter of the trajectories. Holding any mechanical structure would require even more sophisticated means of propulsion.

Using existing space technology it would be possible to construct propulsion systems which could move as the two lights did, but it would require two very sophisticated control systems. Operation of such a system in the Earth's gravity and atmosphere would be several orders of magnitude more difficult than the operation in the space environment. Such a large propulsion and control system would be even more affected by the atmosphere and gravity, requiring an even more complex control system.

Another possible solution would be to use drone technology. However, securing a perfect circular movement, and stabilizing and synchronizing the movement of two drones in the presence of wind and gravity is practically impossible, especially as the two trajectories were very difficult to follow. This solution is rejected because not even the slightest noise was detected during the display. One could draw the conclusion that it would be practically impossible to build such propulsion and control systems using existing technology.

NOTES

Introduction

1. Behe, M.J. 1996. *Darwin's Black Box*. New York: The Free Press.

2. SETI - Search for Extra Terrestrial Intelligence.

3. Ward, P.D., Brownlee, D., 2004, *Rare Earth*. New York: Copernicus Books.

Chapter 1

1. The Drake equation is a probabilistic argument used to arrive at an estimate of the number of active, communicative extraterrestrial civilizations in the Milky Way galaxy.

2. Presented in 2000 by Peter Ward and Donald Brownlee in the book *Rare Earth*.

3. A light year is the distance travelled by light during one year.1 light year = 9.4607×10^{12} km.

4. Hawking radiation is black-body radiation that is predicted to be released by black holes, due to quantum effects near the event horizon.

5. A G class star has about 0.8 to 1.2 solar masses and surface temperature of between 5,300°K and 6,000°K.

6. Depends also on the atmospheric pressure

7. A planet rotating faster would have smaller temperature variations resulting in the reduced circulation rate of the atmosphere.

8. Tilt is measured with reference to the perpendicular to the plane of orbital rotation.

9. When a body is moving with a velocity higher than the escape velocity it will escape into space and never return to the planet.

10. Jeans gas escape velocity is calculated assuming that the mean thermal gas velocity at the upper atmosphere is about one sixth of the planet's escape velocity.

11. It is deducted from seismic tests but it is not known for sure. The deepest drilling has reached about 12 km.

12. The total volume of land above the sea level is 125 million cubic kilometers, less than one tenth of the volume of water.

13. Half life is a time in which half of the elements or chemical compounds are decomposed.

14. Alpha particles consist of helium nuclei.

15. Jupiter's magnetic field is generated by a different mechanism than fields of the terrestrial planets.

16. Ozone consists of three atoms of oxygen, O_3 which absorb 95-97% of UV radiation.

17. AU - the astronomical unit of length, roughly the distance from Earth to the Sun equal to 149.6 million kilometers.

18. Canup, R. M., Asphaug, E. 2001. Origin of the Moon in a giant impact near the end of the Earth's formation. *Nature* 412 (6848): 708–12.
    Stevenson, D. J. 1987. Origin of the moon – The collision hypothesis. *Annual Review of Earth and Planetary Sciences* 15 (1): 271.

Chapter 2

1. Vogler, E.A. 2001. *Biological properties of Water*, in Water in Biomaterials Surface Science, edit. Morra. M. Wiley. www.ems.psu.edu/~vogler/pdfs/Bioprophoh.PDF.

2. Heat capacity is the amount of energy that it takes to raise the temperature of 1 gram of a substance by 1°C.

3. Heat of vaporization is the amount of heat required to convert liquid water into steam.

4. Heat of fusion is the heat you need to take out of water to change it into a solid or the heat you have to put into ice to melt it.

5. This statement might be confusing because water is used for cooling. But in such a case the heat is transferred by the movement of water (convection) not by the transfer of heat through water.

6. The Moon's surface temperature varies over 300°C between day and night.

7. It is a very unusual characteristic that the minimum of water's light absorption spectrum matches the maximum of the Sun's radiation spectrum.

8. Magma ocean event is the stage of a planet's development when high temperature liquid magma is present on the planet's surface.

9. Deuterium is a heavy isotope of hydrogen which contains a neutron in addition to a proton.

10. Hartogh, P., et al. 2011. Ocean-like water in the Jupiter-family comet 03P/Hartley 2. *Nature* 478: 218–20.

11. During hydrogen fusion two isotopes of hydrogen form one atom of helium and generate a large quantity of energy.

12. Saal, A.E., et al. 2013. Hydrogen Isotopes in Lunar Volcanic Glasses and Melt Inclusions Reveal a Carbonaceous Chondrite Heritage. *Science* 340: 1317-1320.

13. The volume of water is about five times larger than the volume of dry land above the sea level.

Chapter 3

1. Viruses are not regarded as living organisms because they need other organisms to support them.

2. Prokaryotes include bacteria and archaea, but for the sake of simplicity we concentrate on bacteria.

3. Hutchinson, C.A., et al. 2016. Design and synthesis of a minimal bacterial genome. *Science.* 351: 25 March.

4. Genes describing DNA polymerase could be mutated, but an organism would not survive such mutations.

5. In yeast an average protein has about 466 amino acids.

6. Averaged mass of atoms used in proteins is about 7 Da.

7. Vendeville, A., et al. 2011. An inventory of the bacterial macromolecular components and their spatial organization, *FEMS Microbiology Review.* http://femsre.oxfordjournals.org/content/35/2/395.

8. McAdams H., Shapiro, L. 2009. System-level design of bacterial cell cycle control. *FEBS Letters* 583 : 3984–3991.

McAdams H., Shapiro, L. 2011. The architecture and conservation pattern of whole cell control circuitry. *J Mol Biol.* May 27; 409(1): 28–35.

9. Components of the control system are defined by DNA, but not its functionality.

Chapter 4

1. Latest research postulate that the eukaryotic cell could have arrived about 2.7 billion years ago. Kazmierczak, J., et al. 2016. Tubular microfossils from 2.8 to 2.7 Ga - old lacustrine deposits of South Africa: A sign for early origin of eukaryotes? *Precambrian Research.* 286: 180–194.

2. Gene expression means that the gene can be activated when it's product is needed or deactivated when it is not needed.

3. Harold, F.M. 2005. Molecules into Cells: Specifying Spatial Architecture. *Microbiology and Molecular Biology Reviews.* Dec.69: 544–564.

4. The protein's structure used for control are recorded in genes, but not the functioning of the control system itself.

Chapter 5

1. Radiation passing through the plane perpendicular to the direction of radiation.

2. Molecular mass frequently called molecular weight is measured in grams/mole.

3. nm – nanometer is one billionth of a meter.

4. Tripathy, B. C., Pattanayak, G. K. 2011. Chlorophyll Biosynthesis in Higher Plants, Chapter Photosynthesis, *Advances in Photosynthesis and Respiration* 34: 63-94.

5. Proton is a hydrogen atom without an electron.

6. Davis, K. M., Pushkar, Y. N. 2015. Structure of the Oxygen Evolving Complex of Photosystem II at Room Temperature. *J. Phys. Chem. B.* 119 (8): 3492–3498.

7. Mass of 1 mole of photosystem II is 700 kg.

8. Molecular complexes are normally built from carbon, hydrogen, oxygen, nitrogen and phosphorus.

9. ATP is generated by ATP synthase described later in this chapter.

10. Bond energy originates from the inter atomic forces holding the molecule together.

11. The Krebs Cycle is also the starting point for the synthesis of the cell's amino acids, fats and other important molecules.

12. Mitchell received the Nobel Prize for chemistry in 1978.

Chapter 6

1. Protista is a group of single eukaryotic cell organisms.

2. Position of land masses is based on the measurements of direction of magnetization of rocks which depends on the direction of the Earth's magnetic field during their formation. The earth's magnetic field has reversed many times and it is not exactly known what direction it had about 600 million years ago.

3. They are subjected to mutations but the mutated animal will die.

4. Concentration gradient is for example, a change of density of molecules along a certain path from high to low.

Chapter 7

1. The plant cuticle is a layer of lipid polymer impregnated with waxes that is present on the outer surfaces of the primary organs of all vascular land plants.

Chapter 8

1. Cranial capacity indicates size of the brain.

2. Cremo, M. A., Thompson, R.L. 1999. *The hidden history of the Human Race.* Los Angeles: Bhaktivedanta book publishing.

3. idem p. 256

4. Krings, M., et all. 1997. Neandertal DNA Sequences and the Origin of Modern Humans. *Cell* 90(1): 19–30.

5. Marean, C.W. 2015. The most invasive species of all, *Scientific American*, August: 23-29.

6. In India and Malaysia were discovered settlements more than 60,000 years old.

Chapter 9

1. Nedergaard, M., Goldman, S.A. 2016. Brain drain. *Scientific American* 314: 44-49.

2. Brain Research through Advancing Innovative Neurotechnologies.

3. http://www.columbia.edu/cu/biology/faculty/yuste/publications.html.

4. The study involved inserting electrodes into the monkey brain.

5. The research was done using rats, but the results can be applied to man as well.

6. Moser, E.I., et al. 2014. Grid cells and cortical representation, *Nature Reviews. Neuroscience* 15: 466-481.

7. Evans, P.D., et al. 2005. Microcephalin, a Gene Regulating Brain Size, Continues to Evolve Adaptively in Humans. *Science* 309: 1717-1720.

8. Abnormal spindle-like micro-cephaly associated.

9. Zika virus is also causing microcephaly.

10. Williams, R. 2006. Genes for bigger brains. *The naked scientists.* April 12.

11. Wilber, K. 1977. *The spectrum of consciousness.* Quest.

12. Jaynes, J., 1976. *The origin of consciousness in the breakdown of the bicameral mind.* New York: Houghton Mifflin.

13. Kulczyk, W.K. 2012. *We are not alone in the Universe.* Winchester: John Hunt Publishing.

Chapter 10

1. Behe, M.J. 2007. *The Edge of Evolution.* New York: Free Press. p. 4.

2. Harold, F. M. 2014. *In search of cell history*. Chicago: The Univ. of Chicago Press, p. 213.

3. Central dogma states that information can flow from DNA to proteins but not in the opposite direction

4. Behe. 2007. p.5

5. Sanjuán, R., Moya, A., Elena, S.F. 2004. The distribution of fitness effects caused by single-nucleotide substitutions in an RNA virus. *Proc Natl Acad Sci U S A* 101(22): 8396–8401.

6. It is assume that mutation rate is 2.5×10^{-8} and is constant across the animal world.

7. Behe. 2007. p. 57.

8. Behe. 2007. p. 61.

9. In animal reproduction the children's cells are copies of the mother's cell.

10. A transposon is a DNA sequence that can change its position within a genome, sometimes creating or reversing mutations and altering the cell's genome size.

Chapter 11

1. Blue Whale has about 100 quadrillions cells (10^{17}).

2. The wheat genome has about 16 billion base pairs.

3. Behe. 1996.

4. Idem. p. 22.

5. Behe. 2007.

6. Balakirev, E.S., Ayala, F.J. 2003. Pseudogenes: Are They "Junk" or Functional DNA? *Annu. Rev. Genet.* 37:123–51.

7. DNA is packed quite tightly to fit into cells, so it has to be partially unfolded to makes genes accessible and allow them to be activated.

8. Jacques, P., et all. 2013. The Majority of Primate-Specific Regulatory Sequences Are Derived from Transposable Elements. *PLoS Genetics* 9(5) doi:10.1371/journal.pgen.1003504.

9. Enard, D., at all. 2016. Viruses are a dominant driver of protein adaptation in mammals. *eLife* 5. DOI: 10.7554/eLife.12469.

10. Feschotte, C., Pritham, E.J. 2007. DNA Transposons and the Evolution of Eukaryotic Genomes. *Annu Rev Genet.* 41: 331–368.

11. Eldredge, N., Gould, S.J. 1972. Punctuated equilibria: an alternative to phyletic gradualism. In T.J.M. Schopf, ed. *Models in Paleobiology* (pp. 82-115). San Francisco: Freeman Cooper.

12. McClintock, B. 1984. The significances of responses of the genome to challenge. *Science* 226: 792– 801.

13. Wong, K. 2000. High-Speed Speciation. *Scientific American*, October 20.

14. Recombination is a process by which pieces of DNA are broken and recombined to produce new combinations of alleles. This recombination process creates genetic diversity at the level of genes that reflects differences in the DNA sequences of different organisms.

Chapter 12

1. Forces between molecules are most effective at distances of about 0.3 nm. Strength of the forces acting between molecules diminishes very rapidly, even as much as with the sixth power of distance. For example, doubling the distance would reduce the strength of molecular forces 64 times.

2. McAdams, H. H., Shapiro, L. 2009. System-level design of bacterial cell cycle control, *FEBS Letters*. 583: 3984–3991.

3. Stem cells are original not differentiated cells during the early development of embryo.

Chapter 13

1. Kelvin scale starts at absolute zero which is -273° C. One degree Kelvin is equal to one degree Centigrade.

2. Last Universal Common Ancestor.

3. Some estimates are up to 1 million species.

4. Dehal, P., Boore, J.L. 2005. Two Rounds of Whole Genome Duplication in the Ancestral Vertebrate. *PLoS Biol* 3(10): e314.

5. Enard, D., et all. 2016. Viruses are a dominant driver of protein adaptation in mammals. *eLife* 5. DOI: 10.7554/eLife.12469.

6. Magiorkinis, G., et all. 2015. The decline of human endogenous retroviruses: extinction and survival. *Retrovirology* 12: 8.

7. Wildschuttea, J. H., et all. 2016. Discovery of unfixed endogenous retrovirus insertions in diverse human populations. *PNAS* 113: E2326–E2334.

8. Zhang, Y.E., et all. 2011. Accelerated Recruitment of New Brain Development Genes into the Human Genome. *PLoS Biol* 9(10): e1001179.

9. Gokhman, D., et all. 2014. Reconstructing the DNA Methylation Maps of the Neandertal and the Denisovan. *Science* 344 (6183): 523-527.

10. Green, R. E., et al. 2010. A draft sequence of the Neandertal genome. *Science* 328: 710-722.

11. Evans, P.D., et al. 2005. Microcephalin, a Gene Regulating Brain Size, Continues to Evolve Adaptively in Humans. *Science* 309: 1717-1720.

12. Catal Huyuk was a very large Neolithic proto-city settlement in southern Anatolia in present Turkey, which existed from approximately 7500 BC to 5700 BC.

13. Kulczyk. 2012.

14. Cremo, M. A., Thompson, R.L. 1999. *The hidden history of the Human Race*. Los Angeles: Bhaktivedanta book publishing.

15. Crick, F.H.C., Orgel, L. E. 1973. Directed panspermia. *Icarus* 19: 341-346.

16. Hoyle, F., Wickramasinghe, N.C. 1980. *Evolution From Space*. London: J.M. Dent.

17. Shcherbak, V.I,, Makukov, M.A. 2013. The "Wow! signal" of the terrestrial genetic code. *Icarus* 224: 228–242.

18. Kulczyk. 2012. p.264

19. Kahneman, D. 2011. *Thinking, Fast and Slow*. Penguin Books.

Chapter 14

1. Search for Extra Terrestrial Intelligence.

2. In July 2015 physicist Stephen Hawking and investor Yuri Milner lunched $100 million initiative called *Breakthrough listen*.

3. Kulczyk. 2012.

4. Ibidem.

5. The project Blue Book was created in 1952 by the US government and had the task of dispelling doubts associated with the observation of UFOs and to explain their origin. Archives of the project contained more than 13,000 reports of unusual events.

6. Piechota, K. 1991. Nocna parada krzyzakow. *Magazyn ufologiczny UFO*. 4(8): 26-46

Appendix 1

1. Walter Hohmann, The Attainability of Heavenly Bodies (Washington: NASA Technical Translation F-44, 1960).

2. http://lasp.colorado.edu/~bagenal/3720/CLASS6/6Equilibrium Temp.html.

Appendix 2

1. Piechota. 1991.

GLOSSARY

amino acid – one of 20 molecular building blocks that are linked together in a chain to form a protein. Their molecular mass* is between 75 and 250 Daltons*

ADP - adenosine diphosphate is the breakdown product of ATP*. It is re-used and changed back to ATP when a phosphate group is added. Its chemical formula is $C_{10}H_{15}N_5O_{10}P_2$.

antibodies - proteins that are produced by the immune system to stop intruders such as bacteria and viruses from harming the body.

astrocytes - the most numerous cell type within the central nervous system which perform a variety of supporting tasks, from axon guidance and synaptic support, to the control of the blood brain barrier and blood flow. They surround neurons in the brain and spinal cord, outnumbering neurons 50:1.

ATP - adenosine triphosphate is the biological energy fuel used by all known cells. Its chemical formula is $C_{10}H_{16}N_5O_{13}P_3$, Its molecular mass is 507 Da. It is made by ATP synthase from ADP by adding one phosphate group.

ATP synthase - an ingenious rotating nano-motor that sits in the membrane and uses the flow of protons (H^+) to power the synthesis of ATP*.

bacteria - simple organisms, size about 1-10 micrometers, existing for at least 3.7 billion years. Bacteria are prokaryotes, which lack nucleus to store their DNA. It is the most abundant organism on Earth

Calvin cycle - a series of biochemical reactions that take place in the photosynthetic organisms. The reactions of the Calvin cycle use carbon from carbon dioxide in the atmosphere and energy from ATP to produce glucose.

catalyst – a substance which increases many times the rate of chemical reactions. In the cell enzymes work as catalysts.

central dogma of molecular biology describes the two-step process by which genetic information flows into proteins: DNA → RNA → protein. It states that such information cannot be transferred back from protein to DNA.

centriole - a small cylindrical organelle that helps cells divide, or make copies of themselves. All centrioles are made of protein strands called microtubules.

cerebral cortex – the outermost layered structure of the brain which controls higher brain functions such as information processing. The cerebral cortex is composed of gray matter, consisting mainly of cell bodies with astrocytes* being the most abundant cell type.

cerebellum – the part of the brain at the back of the skull in vertebrates where the spinal cord meets the brain. The cerebellum coordinates voluntary movements such as posture, balance, coordination and speech. It contains roughly half of the brain's neurons.

cerebrospinal fluid - is a clear, colorless body fluid found in the brain and spinal cord. It acts as a cushion or buffer for the brain, providing basic mechanical and immunological protection to the brain inside the skull.

chaperon – a special protein that assist the folding or unfolding and the assembly or disassembly of macromolecular structures such as proteins.

chloroplast - a specialized component in plant cells and algae where photosynthesis take place. It contains chlorophyll – a green pigment.

chlorophyll – a green pigment found in cyanobacteria and the chloroplasts of algae and plants. It is an extremely important biomolecule, critical in photosynthesis, which allows plants to absorb energy from light.

chromosome – a tubular structure composed of DNA tightly wrapped in protein. It contains several genes. Humans have 23 pairs of chromosomes containing two copies of all our genes

cofactor – a non protein compound made of amino-acids that assist in biochemical transformations. It is involved in activating enzymes.

Corti organ - the receptor organ for hearing and is located in the mammalian cochlea. It contains the hair cells that give rise to nerve signals in response to sound vibrations.

cortical areas – parts of the cerebral cortex performing specific functions originally defined by Brodmann.

cyanobacteria - a group of photosynthetic, nitrogen fixing bacteria that live in a wide variety of habitats such as moist soils and in water. Cyanobacteria have a unique set of pigments used in photosynthesis which can give some of them a blue-green color. They are the oldest known fossils, more than 3.5 billion years old.

cytochromes - a class of proteins that transport electrons or protons during cell respiration or photosynthesis. Cytochromes are found in the mitochondrial inner membrane of eukaryotes, in the chloroplasts of plants, in photosynthetic microorganisms, as well as in bacteria.

cytoplasm – part of the cells outside nucleus consisting of watery solution and some cell components such as mitochondria.

cytoskeleton – a microscopic network of protein filaments and tubules in the cytoplasm of many living cells, giving them shape and coherence.

Dalton – a unit of molecular mass defined as 1/12 weight of carbon-12 atom in the ground state.. A unit of one Dalton (abr. Da) corresponds to the mass of one 1 g/mole, equal to mass of one hydrogen nucleus or 1.67 x 10^{-24}grams. A thousand Daltons is abbreviated as kDa.

DNA – deoxyribonucleic acid carries genetic information and is present in every cell of the organism. Its two strands are made from nucleotides* linked together and forming a double helix.

DNA polymerase – a complex molecule which makes copies of DNA during the reproductive cycle. It is a very large molecule and in E. coli* is formed from 928 amino acids* and its consists of about 19 thousand atoms.

E. coli, *Escherichia coli* - a rod-shaped bacteria living in the large intestine of humans and other animals. Frequently used in biological research.

electron transport chain - the final stage of respiration leading to the forming of ATP in the inner membrane of the mitochondrion. It comprises of five molecular complexes which use electron energy to pump protons* across the membrane. The last complex is ATP synthase.

endocytosis - a form of active transport in which a cell transports molecules into the cell by engulfing them in an energy-using process.

endoplasmic reticulum- a type of organelle in the eukaryotic cells that forms an interconnected network of flattened, membrane-enclosed sacs or tube-like structures. It functions as the transportation system of the eukaryotic cell, and its proteins are contained within it until they are needed to move.

enzyme – a protein that is used to catalyze a particular chemical reaction increasing its rate up to by millions of times the normal rate.

eukaryote- an organism composed of one or more cells containing a nucleus and other specialized structures like mitochondria and cytoskeleton. All complex life such as plants and animals are made of eukaryotic cells.

ganglion cell - a retinal ganglion cell is a type of neuron located near the inner surface of the retina of the eye. It receives visual information from photoreceptors via two intermediate neuron cells and generates signal which is sent to the brain.

gene - a unit of heredity, usually consisting of a stretch of DNA encoding a protein.

glial cells - surround neurons, provide support and insulate one neuron from another. They also supply nutrients and oxygen to neurons.

Golgi apparatus - a series of membranes shaped like pancakes. The membrane surrounds an area of fluid where the complex molecules such as proteins, sugars, enzymes are stored and changed. The Golgi apparatus is responsible for transporting, modifying, and packaging proteins and lipids into vesicles for delivery to targeted destinations

hippocampus - a small formation in the brain that plays an important role in the formation of new memories and is also associated with learning and emotions.

horizontal cells - are the laterally interconnecting neurons in the inner nuclear layer of the retina of vertebrate eyes.

hydrophilic - molecules that are attracted and dissolved by water

hydrophobic - molecules that repel water molecules

inferior colliculus - are responsible for processing the signals derived from the left and right ears

intron – a sequence of DNA within a gene which does not code for a protein and is usually removed from the gene before the protein is made.

Krebs cycle or citric acid cycle - the sequence of reactions by which most living cells generate energy during the process of respiration. It takes place in the mitochondria, using up oxygen and producing carbon dioxide and water as waste products.

lateral geniculate nucleus, LGN - a relay centre in the thalamus for the visual pathway. It receives a major sensory input from the retina.

lipids – a long chain hydrocarbon used in membranes of bacteria and eukaryotes

LUCA – the last universal common ancestor of all cells living today whose hypothetical properties could be estimated.

lymph - fluid that removes excess plasma, dead blood cells, debris and other waste from the body.

Magnetic Resonance Imaging, MRI – a type of scan that uses strong magnetic fields, radio waves and a computer to produce detailed images of the inside of the body.

medial geniculate nucleus - a part of the auditory thalamus nuclei which receive, process and then relay auditory information from the ears.

membrane – a very thin layer surrounding cells and found also inside cells composed of lipid bilayer with a hydrophobic interior and hydrophilic exterior on either side.

metabolism – the set of life sustaining chemical reactions within living cells.

metalloproteins - proteins having metal atoms in its structure, often working as a catalyst.

mitochondrion - a separate part in eukaryotic cells which generate energy from sugar by making ATP molecules. Mitochondria include en electron transport chain.

molecular mass - a number equal to the sum of the atomic masses of the atoms in a molecule. See Dalton.

molecular clock - a technique that uses the mutation rate of biomolecules to deduce the time in prehistory when two or more life forms diverged. It looks into the changes in the amino acid sequences of proteins that took place during this time.

mRNA, messenger RNA - a single-stranded molecule of RNA* that is synthesized from a DNA template where its genetic code specifies the amino acid sequence for protein synthesis.

mutation – a change in the specific sequence of nucleotides of a gene, but also includes random deletions or duplication of DNA.

NADPH, *nicotinamide adenine dinucleotide phosphate* - an organic molecule that is produced in plants during photosynthesis Their main job in the cell is to carry electrons and protons around.

neuron - a nerve cell that processes and transmits information through electrical and chemical signals.

nucleic acids - are biopolymers or large biomolecules, essential for all known forms of life. Nucleic acids, which include DNA (deoxyribonucleic acid) and RNA (ribonucleic acid), are made from nucleotides.

nucleotide – the basic structural unit and building block for DNA and RNA. These building blocks are hooked together to form a chain of DNA and RNA.

nucleus - a part of eukaryotic cells which contains most of the cell's genes.

optic chiasm - an X-shaped structure formed by the crossing of the optic nerves in the brain. The optic nerve connects the brain to the eye.

osmosis - diffusion of water across a membrane.

oxygen evolving complex – a part of the photosynthesis complex molecule which is involved in water splitting reaction using a catalyst and solar energy.

phospholipids - serve as a major structural component of most biological membranes. They consist of a hydrophilic* head and a hydrophobic tail.

photosynthesis - conversion of carbon dioxide from air into organic matter using solar energy which splits water into hydrogen and oxygen.

polypeptide - a chains of amino acids. Proteins are made up of one or more polypeptide molecules.

promoter - a region of DNA that initiates transcription of a particular gene and provide a secure initial binding site for RNA polymerase. It works as a label identifying a specific gene.

prokaryote – a term describing simple cells including bacteria and archaea that lack a nucleus.

protein – a large molecule, building structure of all cells made of a chain of amino acids specified by the sequence of DNA.

proton – a subatomic particle with positive charge. A proton is a hydrogen atom without an electron.

ribosome – a protein building factory found in all cells. It converts the mRNA* code into a protein using amino acids labeled with tRNA*

RNA - ribonucleic acid made of nucleotides* chains. It is used as a messenger RNA to carry coding from DNA to ribosome and also is used as a building block of many molecular complexes.

RNA polymerase- an enzyme that synthesize RNA* molecule from nucleotides*

rRNA - ribosomal RNA, a molecular component of a ribosome, the cell's essential protein factory. rRNA fabricates the polypeptides and provides a mechanism for decoding mRNA into amino acids and interacts with the tRNA during translation.

tRNA - a transfer RNA is an molecule composed of RNA, typically 76 to 90 nucleotides in length, that serves as a label identifying amino acids to be used by ribosome.

rubisco - ribulose-1,5-bisphosphate carboxylase/oxygenase, commonly known by the abbreviations RuBisCO, is an enzyme involved in the first major step of carbon fixation in the Calvin cycle, a process by which atmospheric carbon dioxide is converted by plants and other photosynthetic organisms to energy-rich molecules such as glucose.

spliceosome - a large and complex molecular machine found primarily within the cell nucleus of eukaryotic cells. The spliceosome removes introns* from a transcribed pre-mRNA* and splice reminding parts of the gene.

synapse – a junction between two nerve cells, consisting of a minute gap across which impulses pass by diffusion of a neurotransmitter.

thalamus – part of the brain with several functions such as relaying of sensory and motor signals to the cerebral cortex, and the regulation of consciousness, sleep, and alertness.

thylakoid - each of a number of flattened sacs inside a chloroplast, bounded by pigmented membranes on which the light reactions of photosynthesis take place.

transcription - the process of making an RNA* copy of a gene sequence. This copy, called a messenger RNA (mRNA) molecule, leaves the cell nucleus and enters the cytoplasm, where it directs the synthesis of the protein, which it encodes. Transcription is carried out by an enzyme called RNA polymerase*.

translation - the process of translating the sequence of a messenger RNA (mRNA) molecule to a sequence of amino acids during protein synthesis. In the cell cytoplasm, the ribosome reads the sequence of the mRNA to assemble the protein.

vacuole - a storage bubble found in cells. It is found in both animal and plant cells but is much larger in plant cells. Vacuoles might store food or any variety of nutrients a cell might need to survive or waste products.

vesicle - a small fluid-filled sac in cells. A membrane-bound sac in eukaryotic cells that stores or transports the products of metabolism in the cell and is sometimes the site for the breaking down of metabolic wastes.

Visual cortex - a part of the cerebral cortex* that processes visual information from the retina. It is located in the back of the head.

LIST OF ILLUSTRATIONS

12. Figure 5-4. Summary of photosynthesis in plants utilizing photosystem I and II.
    Copyright © 2017 Christopher Harley.
13. Figure 5-5. Structure of a mitochondrion.
    Credit: http://cronodon.com/BioTech/Respiration.html. Under the Creative Commons Attribution International license.
14. Figure 5-6. Respiration system in mitochondria.
    Copyright © 2017 Christopher Harley.
15. Figure 5-7. Structure of complex I based on X-ray crystallography.
    Credit: David S. Goodsell.
16. Figure 5-8. Functional diagram of ATP synthase.
    Copyright © 2017 Christopher Harley.
17. Figure 12-1. A diagram showing the principles of negative feedback control.
    Copyright ©2017 W.K. Kulczyk.
18. Figure 12-2. Diagram of the cell control system.
    Copyright ©2017 W.K. Kulczyk.
19. Figure A1-1. Schematic diagram showing Hohmann transfer orbit.
20. Figure A2 -1. Trajectory of lights.
    Copyright ©2017 W.K. Kulczyk.
21. Figure A2- 2. Geometry of the circular motion of the spots of light.
    Copyright ©2017 W.K. Kulczyk.

Printed in Great Britain
by Amazon